THE NEW HARVEST

"This book presents a timely analysis of the importance of infra-structure in improving Africa's agriculture. Leaders at national and state levels will benefit immensely from its evidence-based recommendations."

— *Goodluck Jonathan*
President of the Federal Republic of Nigeria

"This book is a forceful reminder of the important role that African women play in agriculture on the continent. It is critical that they are provided with equal educational opportunity as a starting point for building a new economic future for the continent."

— *Ellen Johnson Sirleaf*
President of the Republic of Liberia

"New technologies, especially biotechnology, provide African countries with additional tools for improving the welfare of farmers. I commend this book for the emphasis it places on the critical role that technological innovation plays in agriculture. The study is a timely handbook for those seeking new ways of harnessing new technologies for development, including poor farmers, many of whom are women."

— *Blaise Compaoré*
President of Burkina Faso

"*The New Harvest* underscores the importance of global learning in Africa's agricultural development. It offers new ideas for international cooperation on sustainable agriculture in the tropics. It will pave the way for improved collaboration between Africa and South America."

— *Laura Chincilla*
President of Costa Rica

THE NEW HARVEST
AGRICULTURAL INNOVATION IN AFRICA

CALESTOUS JUMA

OXFORD

UNIVERSITY PRESS

2011

OXFORD
UNIVERSITY PRESS

Oxford University Press, Inc., publishes works that further
Oxford University's objective of excellence
in research, scholarship, and education.

Oxford   New York
Auckland   Cape Town   Dar es Salaam   Hong Kong   Karachi
Kuala Lumpur   Madrid   Melbourne   Mexico City   Nairobi
New Delhi   Shanghai   Taipei   Toronto

With offices in
Argentina   Austria   Brazil   Chile   Czech Republic   France   Greece
Guatemala   Hungary   Italy   Japan   Poland   Portugal   Singapore
South Korea   Switzerland   Thailand   Turkey   Ukraine   Vietnam

Published by Oxford University Press, Inc.
198 Madison Avenue, New York, NY 10016

www.oup.com

Oxford is a registered trademark of Oxford University Press

Library of Congress Cataloging-in-Publication Data
Juma, Calestous.
The new harvest : agricultural innovation in Africa / Calestous Juma.
p. cm.
Includes bibliographical references and index.
ISBN 978-0-19-978319-9 (pbk.) — ISBN 978-0-19-978320-5 (cloth)
1. Agriculture—Economic aspects—Africa.
2. Agricultural innovations—Economic aspects—Africa.
3. Economic development—Africa. I. Title.
HD9017.A2J86   2011
338.1'6096—dc22
2010030674

1 3 5 7 9 8 6 4 2

Printed in the United States of America
on acid-free paper

In memory of Christopher Freeman
Father of the field of science policy and innovation studies

CONTENTS

Agricultural Innovation in Africa Project

Project Director and Lead Author
> Calestous Juma, Harvard Kennedy School, Harvard University, Cambridge, USA

International Advisory Panel and Contributing Authors
> John Adeoti, Nigerian Institute of Social and Economic Research, Ibadan, Nigeria
>
> Aggrey Ambali, NEPAD Planning and Coordinating Agency, Tshwane, South Africa
>
> N'Dri Assié-Lumumba, Cornell University, Ithaca, USA
>
> Zhangliang Chen, Guangxi Zhuang Autonomous Region, Nanning, People's Republic of China
>
> Mateja Dermastia, Anteja ECG, Ljubljana, Slovenia
>
> Anil Gupta, Society for Research and Initiatives for Sustainable Technologies and Institutions, and Indian Institute of Management, Ahmedabad, India
>
> Daniel Kammen, University of California, Berkeley, USA
>
> Margaret Kilo, African Development Bank, Tunis, Tunisia
>
> Hiroyuki Kubota, Japan International Cooperation Agency, Tokyo
>
> Francis Mangeni, Common Market for Eastern and Southern Africa, Lusaka, Zambia
>
> Magdy Madkour, Ain Shams University, Cairo, Egypt
>
> Venkatesh Narayanamurti, School of Engineering and Applied Sciences, Harvard University, USA
>
> Robert Paarlberg, Wellesley College and Harvard University, USA
>
> Maria Jose Sampaio, Brazilian Agricultural Research Corporation, Brasilia, Brazil
>
> Lindiwe Majele Sibanda, Food, Agriculture and Natural Resources Policy Analysis Network, Tshwane, South Africa
>
> Greet Smets, Biotechnology and Regulatory Specialist, Essen, Belgium
>
> Botlhale Tema, African Creative Connections, Johannesburg, South Africa
>
> Jeff Waage, London International Development Centre, London
>
> Judi Wakhungu, African Centre for Technology Studies, Nairobi, Kenya

Project Coordinator
> Greg Durham, Harvard Kennedy School, Harvard University, Cambridge, USA

ACKNOWLEDGMENTS

This book is a product of the Agricultural Innovation in Africa (AIA) project funded by the Bill and Melinda Gates Foundation. Its production benefited greatly from the intellectual guidance and written contributions of the members of the AIA International Advisory Panel. In addition to providing their own contributions, the panel members have played a critical role in disseminating AIA's preliminary findings in key policy and scholarly circles. The production of this book would not have been possible without their genuine support for the project and dedication to the cause of improving African agriculture.

We are deeply grateful to the information provided to us by Juma Mwapachu (Secretary General), Alloys Mutabingwa, Jean Claude Nsengiyumva, Phil Klerruu, Moses Marwa, Flora Msonda, John Mungai, Weggoro Nyamajeje, Henry Obbo, and Richard Owora at the Secretariat of the East African Community (Arusha, Tanzania) on Africa's Regional Economic Communities (RECs). Sam Kanyarukiga, Frank Mugyenyi, Martha Byanyima, Jan Joost Nijhoff, Nalishebo Meebelo, and Angelo Daka (Common Market for Eastern and Southern Africa, Zambia) provided additional contributions on regional perspectives. We have benefited from critical insights from Nina Fedoroff and Andrew Reynolds (U.S. Department of

State, Washington, DC), Andrew Daudi (Office of the President, Lilongwe, Malawi), Norman Clark (African Centre for Technology Studies, Nairobi, Kenya), Daniel Dalohoun (AfricaRice, Cotonou, Benin), Andy Hall (UNU-MERIT, Maastricht, the Netherlands), and Peter Wanyama (Mohammed Muigai Advocates, Nairobi, Kenya). Additional information for the book was provided by John Pasek (Iowa State University, USA), Stephanie Hanson (One Acre Fund, Bungoma, Kenya), S. G. Vombatkere (Mysore, India), Akwasi Asamoah (Kwame Nkrumah University of Science and Technology, Kumasi, Ghana), Jonathan Gressel (Weizmann Institute and Trans-Algae Ltd., Israel), Will Masters (Tufts University, USA), Martha Dolpen (African Food and Peace Foundation, USA), and C. Ford Runge (University of Minnesota, USA).

We would like to thank our dedicated team of researchers who included Edmundo Barros, Daniel Coutinho, Manisha Dookhony, Josh Drake, Emily Janoch, Beth Maclin, Julia Mensah, Shino Saruti, Mahat Somane, and Melanie Vant at the Harvard Kennedy School, and Yunan Jin, Jeff Solnet, Amaka Uzoh, and Caroline Wu at Harvard College, Harvard University.

Much inspiration for the structure of this book came from the results of a meeting of experts, "Innovating Out of Poverty," convened by the Organisation for Economic Co-operation and Development (OECD) in Paris on April 6–7, 2009. It gathered state-of-the-art knowledge on agricultural innovation systems. We are particularly grateful to Richard Carey, Kaori Miyamoto, and Fred Gault (OECD Development Centre, Paris). Further input was provided by David Angell (Ministry of Foreign Affairs, Ottawa), Jajeev Chawla (Survey Settlements and Land Records Department, India), David King (International Federation of Agricultural Producers, Paris), Raul Montemayor (Federation of Free Farmers, Manila), Charles Gore (United Nations Conference on Trade and Development, Geneva), Paulo Gomes (Constelor Group, Washington, DC), Ren Wang (Consultative Group on International Agricultural Research, Washington,

DC), David Birch (Consult Hyperion, London), Laurens van Veldhutzen (ETC Foundation, the Netherlands), Erika Kraemer-Mbula (Centre for Research in Innovation Management, Brighton University, UK), Khalid El-Harizi (International Fund for Agricultural Development, Rome), Adrian Ely (University of Sussex, UK), Watu Wamae (The Open University, UK), Andrew Hall (United Nations University, the Netherlands), Rajendra Ranganathan (Utexrwa, Kigali), Alfred Watkins (World Bank, Washington, DC), Eija Pehu (World Bank, Washington, DC), and Wacege Mugua (Safaricom, Ltd., Nairobi).

Our ideas about the role of experiential learning has been enriched by the generosity of José Zaglul and Daniel Sherrard at EARTH University and their readiness to share important lessons from their experience running the world's first "sustainable agriculture university." We drew additional inspiration and encouragement from Alice Amsden (Massachusetts Institute of Technology, USA), Thomas Burke (Harvard Medical School, USA), Gordon Conway (Imperial College, London), David King (University of Oxford, UK), Yee-Cheong Lee (Academy of Sciences, Malaysia), Peter Raven (Missouri Botanical Garden, USA), Ismail Serageldin (Library of Alexandria, Egypt), Gus Speth (Vermont Law School, USA), and Yolanda Kakabadse (WWF International, Switzerland).

I am grateful to the Bill and Melinda Gates Foundation and in particular to Gwen Young, Sam Dryden, Brantley Browning, Lawrence Kent, Corinne Self, Prabhu Pingali, and Greg Traxler for continuing support to the Agricultural Innovation in Africa Project at the Belfer Center for Science and International Affairs of the Harvard Kennedy School. I am also indebted to my colleagues at the Harvard Kennedy School who have given me considerable moral and intellectual support in the implementation of this initiative. We want to single out Graham Allison, Venkatesh Narayanamurti, Bill Clark, Nancy Dickson, Eric Rosenbach, and Kevin Ryan for their continuing support.

Special credit goes to Greg Durham (AIA Project Coordinator, Harvard Kennedy School) who provided invaluable managerial support and additional research assistance. Finally, we want to thank Angela Chnapko and Joellyn Ausanka (Oxford University Press) and anonymous reviewers for their constructive editorial support and guidance.

New ideas need champions. We would like to commend Sindiso Ngwenya (Secretary General of the Common Market for Eastern and Southern Africa, Lusaka) for taking an early lead to present the ideas contained in this book to African heads of state and government for consideration. The results of his efforts are contained in Appendix II. We hope that others will follow his example.

INTRODUCTION

In his acceptance speech as Chairman of the Assembly of the African Union (AU) in February 2010, President Bingu wa Mutharika of Malawi said:

> One challenge we all face is poverty, hunger and malnutrition of large populations. Therefore achieving food security at the African level should be able to address these problems. I would therefore request the AU Assembly to share the dream that five years from now no child in Africa should die of hunger and malnutrition. No child should go to bed hungry. I realize that this is an ambitious dream but one that can be realized. We all know that Africa is endowed with vast fertile soils, favourable climates, vast water basins and perennial rivers that could be utilized for irrigation farming and lead to the Green Revolution, and mitigate the adverse effects of climate change. We can therefore grow enough food to feed everyone in Africa. I am, therefore, proposing that our agenda for Africa should focus on agriculture and food security. I propose that our slogan should be: "Feeding Africa through New Technologies: Let Us Act Now."[1]

This statement lays out a clear vision of how to approach Africa's agricultural challenge. This book builds on this optimistic outlook against a general background of gloom that fails to account for a wide range of success stories across the continent.[2] African agriculture is at the crossroads. Persistent food shortages are now being compounded by new threats arising from climate change. But Africa faces three major opportunities that can help transform its agriculture to be a force for economic growth. First, advances in science, technology, and engineering worldwide offer Africa new tools needed to promote sustainable agriculture. Second, efforts to create regional markets will provide new incentives for agricultural production and trade. Third, a new generation of African leaders is helping the continent to focus on long-term economic transformation. This book provides policy-relevant information on how to align science, technology, and engineering missions with regional agricultural development goals.[3]

This book argues that sustaining African economic prosperity will require significant efforts to modernize the continent's economy through the application of science and technology in agriculture. In other words, agriculture needs to be viewed as a knowledge-based entrepreneurial activity.[4] The argument is based on the premise that smart investments in agriculture will have multiplier effects in many sectors of the economy and help spread prosperity. More specifically, the book focuses on the importance of boosting support for agricultural research as part of a larger agenda to promote innovation, invest in enabling infrastructure, build human capacity, stimulate entrepreneurship and improve the governance of innovation.

The emergence of Africa's Regional Economic Communities (RECs) provides a unique opportunity to promote innovation in African agriculture in a more systematic and coordinated way.[5] The launching of the East African Common Market in July 2010 represented a significant milestone in the steady process of deepening Africa's economic integration. It is a trend that complements similar efforts in other parts of Africa. It also underscores the determination among African leaders to expand prospects for prosperity by creating space for economic growth and technological innovation.

One of the challenges facing Africa's RECs has been their perceived overlap and duplication of effort. Part of this concern has been overstated. The RECs evolved based on local priorities. For example, the Economic Community of West African States (ECOWAS) is by far the most advanced in peace-keeping while the Common Market for Eastern and Southern Africa (COMESA) has made significant strides in trade matters. In the meantime, the East African Community (EAC)—one of the oldest regional integration bodies in the world—has made significant advances on the social, cultural, and political fronts. It has judicial and legislative organs and aspires to create a federated state with a single president in the future.

Probably the most creative response to concerns over overlap and duplication was the 2008 communiqué of October 22 by the heads of state and government of COMESA, EAC, and the Southern African Development Community (SADC), agreeing to form a tripartite free trade area covering the 26 countries in the region, and to cooperate in various areas with SADC. This move will go a long way in helping achieve the African Union's continental integration objectives. The agreement provides for free movement of businesspeople, joint implementation of inter-regional infrastructure programs, and institutional arrangements that promote cooperation among the three RECs. The agreement calls for immediate steps to merge the three trading blocs into a single REC with a focus on fast-tracking the creation of the African Economic Community.

Drawing on the experience of the EAC, the Tripartite Task Force, made up of the secretariats of the three RECs, will develop a road map for the implementation of the merger. The new trading bloc will have a combined GDP of US$625 billion and a population of 527 million. It is estimated that exports among the 26 countries rose from US$7 billion in 2000 to US$27 billion in 2008, and imports jumped from US$9 billion in 2000 to US$32 billion in 2008. This impressive increase was attributed to the efforts of the three RECs to promote free trade.

The heads of state and government directed the three RECs to implement joint programs: a single airspace; an accelerated, seamless inter-regional broadband infrastructure network; and a harmonized policy and regulatory framework to govern information and communication technology (ICT) and infrastructure development. The RECs were also expected to effectively coordinate and harmonize their transport, energy, and investment master plans. The three secretariats were asked to prepare a joint financing and implementation mechanism for infrastructure development within a year. It was also agreed that a Tripartite Summit of heads of state and government shall meet once every two years.

There are other regional developments in Africa that project a different scenario. The Euro-Mediterranean Partnership with the Maghreb countries (Algeria, Morocco, and Tunisia) resembles hub-and-spoke bilateral networks. The agreements signaled interest in facilitating free trade and promoting foreign direct investment. They also represented innovations in international cooperation that defied classical "north-south" relations.[6] While the arrangements have helped to safeguard access by the Maghreb countries to the European Union (EU) countries, they have yet to show strong evidence of foreign direction investment (FDI) flows.[7] A more generous interpretation is that FDI flows are more complex to arrange outside the general domain of natural resource extraction. It may take time for such flows to occur, and some of the decisions might be linked to access critical resources such as solar energy from the Sahara Desert. This topic continues to attract attention in business and technical circles.[8]

It would appear for the time being that regional integration in eastern and southern Africa is driven more by regional trade dynamics while the Maghreb has to endure the pressures of being close to the European Union, with attendant uncertainties on the extent to which Euro-Mediterranean relations can maintain a dependable path.[9] A series of missteps and false starts in EU integration have resulted in a more precautionary approach that is needed to foster policy learning. These dynamics are likely to influence the types of technological trajectories that the various regions of Africa pursue.

The Economic Community of West African States, for example, might end up developing new southern transatlantic trade relations with the United States and South America while eastern and southern Africa may turn east. These scenarios will influence the kinds of technologies and strategies that the regions adopt.

This book builds on the findings of the report, *Freedom to Innovate: Biotechnology in Africa's Development*, prepared by the High Level African Panel on Modern Biotechnology of the African Union (AU) and the New Partnership for Africa's Development (NEPAD).[10] The panel's main recommendations include the need for individual countries in central, eastern, western, northern, and southern Africa to work together at the regional level to scale up the development of biotechnology. This book aims to provide ideas on how to position agriculture at the center of efforts to spur economic development in Africa. It outlines the policies and institutional changes needed to promote agricultural innovation in light of changing ecological, economic, and political circumstances in Africa.

This book explores the role of rapid technological innovation in fostering sustainability, with specific emphasis on sustainable agriculture. It provides illustrations from advances in information technology, biotechnology, and nanotechnology. It builds on recent advances in knowledge on the origin and evolution of technological systems. Agricultural productivity, entrepreneurship, and value addition foster productivity in rural-based economies. In many poor countries, however, farmers, small and medium-sized enterprises, and research centers do not interact in ways that accelerate the move beyond low value-added subsistence sustainable agriculture. Strengthening rural innovation systems, developing effective clusters that can add value to unprocessed raw materials, and promoting value chains across such diverse sectors as horticulture, food processing and packaging, food storage and transportation, food safety, distribution systems, and exports are all central to moving beyond subsistence sustainable agriculture, generating growth, and moving toward prosperity.

Developed and emerging economies can do much more to identify and support policies and programs to assist Africa in taking a comprehensive approach to agricultural development to break out of poverty. This requires rethinking the agenda to create innovation systems to foster interactions among government, industry, academia, and civil society—all of which are critical actors.

The book is guided by the view that innovation is the engine of social and economic development in general and agriculture in particular. The current concerns over rising food prices have compounded concerns about the state and future of African agriculture. This sector has historically lagged behind the rest of the world. Part of the problem lies in the low level of investment in Africa's agricultural research and development. Enhancing African agricultural development will require specific efforts aimed at aligning science and technology strategies with agricultural development efforts. Furthermore, such efforts will need to be pursued as part of Africa's growing interest in regional economic integration through its Regional Economic Communities.

African leaders have in recent years been placing increasing emphasis on the role of science and innovation in economic transformation. The Eighth African Union Summit met in

January 2007 and adopted decisions aimed at encouraging more African youth to take up studies in science, technology, and engineering education; promoting and supporting research and innovation activities and the related human and institutional capacities; ensuring scrupulous application of scientific ethics; revitalizing African universities and other African institutions of higher education as well as scientific research institutions; promoting and enhancing regional as well as south-south and north-south cooperation in science and technology; increasing funding for national, regional, and continental programs for science and technology; and supporting the establishment of national and regional centers of excellence in science and technology.[11] The decisions are part of a growing body of guidance on the role of science and innovation in Africa's economic transformation. These decisions underscore the growing importance that African leaders place on science and innovation for development.

However, the translation of these decisions into concrete action remains a key challenge for Africa. This study is guided by the view that one of the main problems facing African countries is aligning national and regional levels of governance with long-term technological considerations. This challenge is emerging at a time when African countries are seeking to deepen economic integration and expand domestic markets. These efforts are likely to affect the way agricultural policy is pursued in Africa.

The 2007 African Union Summit decisions paid particular attention to the role of science, technology, and innovation in Africa's economic transformation, and they marked the start of identifying and building constituencies for fostering science, technology, and innovation in Africa. They focused on the need to undertake the policy reforms necessary to align the missions and operations of institutions of higher learning with economic development goals in general and the improvement of human welfare in particular.

These decisions represent a clear expression of political will and interest in pursuing specific reforms that would help in making science, technology, and innovation relevant to development. However, the capacity to do so is limited by the lack of informed advice on international comparative experiences on the subject. The central focus of this book is to provide high-level decision makers in Africa with information on how to integrate science and technology into agricultural development discussions and strategies. Specific attention will be placed on identifying emerging technologies and exploring how they can be adapted to local economic conditions.

The book is divided into seven chapters. Chapter 1 examines the critical linkages between agriculture and economic growth. The current global economic crisis, rising food prices, and the threat of climate change have reinforced the urgency to find lasting solutions to Africa's agricultural challenges. The entire world needs to find ways to intensify agricultural production while protecting the environment.[12] Africa is largely an agricultural economy, with the majority of the population deriving their income from farming. Food security, agricultural development, and economic growth are intertwined. Improving Africa's agricultural performance will require deliberate policy efforts to bring higher technical education, especially in universities, to the service of agriculture and the economy. It is important to focus on how to improve the productivity of agricultural workers, most of whom are women, through technological innovation.

Chapter 2 reviews the implications of advances in science and technology for Africa's agriculture. The Green Revolution played a critical role in helping to overcome chronic food shortages in Latin America and Asia. The Green Revolution was largely a result of the creation of new institutional arrangements aimed at using existing technology to improve agricultural productivity. African countries are faced with enormous technological challenges. But they also have access to a much larger pool of scientific and technical knowledge than was available when the Green Revolution was launched. It is important to review major advances in science, technology, and engineering and identify their potential for use in African agriculture. Such exploration should include an examination of local innovations as well as indigenous knowledge. It should cover fields such as information and communication technology, genetics, ecology, and geographical sciences. Understanding the convergence of these and other fields and their implications for African agriculture is important for effective decision making and practical action.

Chapter 3 provides a conceptual framework for defining agricultural innovation in a systemic context. The use of emerging technology and indigenous knowledge to promote sustainable agriculture will require adjustments in existing institutions. New approaches will need to be adopted to promote close interactions between government, business, farmers, academia, and civil society. It is important to identify novel agricultural innovation systems of relevance to Africa. This chapter examines the connections between agricultural innovation and wider economic policies. Agriculture is inherently a place-based activity and so the book outlines strategies that reflect local innovation clusters and other characteristics of local innovation systems. Positioning sustainable agriculture as a knowledge-intensive sector will require fundamental reforms in existing learning institutions, especially universities and research institutes. Most specifically, key functions such as research, teaching, extension, and commercialization need to be much more closely integrated.

In Chapter 4 the book outlines the critical linkages between infrastructure and agricultural innovation. Enabling infrastructure (covering public utilities, public works, transportation, and research facilities) is essential for agricultural development. Infrastructure is defined here as facilities, structures, associated equipment, services, and institutional arrangements that facilitate the flow of agricultural goods, services, and ideas. Infrastructure represents a foundational base for applying technical knowledge in sustainable development and relies heavily on civil engineering. The importance of providing an enabling infrastructure for agricultural development cannot be overstated. Modern infrastructure facilities will also need to reflect the growing concern over climate change. In this respect, the chapter will focus on ways to design "smart infrastructure" that takes advantage of advances in the engineering sciences as well as ecologically sound systems design. Unlike other regions of the world, Africa's poor infrastructure represents a unique opportunity to adopt new approaches in the design and implementation of infrastructure facilities.

The role of education in fostering agricultural innovation is the subject of Chapter 5. Some of Africa's most persistent agricultural challenges lie in the educational system. Much of the focus of the educational system is training young people to seek employment in urban areas. Much of the research is carried out in institutions that do not teach, while universities have limited access to research support. But there is an urgency to identify new ways to enhance competence throughout the agricultural value chain, with emphasis on the role of women as farm workers and custodians of the environment. It is important to take a pragmatic approach that emphasizes competence building as a key way to advance social justice. Most of the strategies to strengthen the technical competence of African farmers will entail major reforms in existing universities and research institutions. In this respect, actions need to be considered in the context of agricultural innovation systems.

Chapter 6 presents the importance of entrepreneurship in agricultural innovation. The creation of agricultural enterprises represents one of the most effective ways to stimulate rural development. The chapter will review the efficacy of the policy tools used to promote agricultural enterprises. These include direct financing, matching grants, taxation policies, government or public procurement policies, and rewards to recognize creativity and innovation. It is important to learn from examples that helped to popularize modern technology in rural areas and has spread to more than 90% of the country's counties. Inspired by such examples, Africa should explore ways to create incentives that stimulate entrepreneurship in the agricultural sector. It is important to take into account new tools such as information and communication technologies and the extent to which they can be harnessed to promote entrepreneurship.

The final chapter outlines regional approaches for fostering agricultural innovation. African countries are increasingly focusing on promoting regional economic integration as a way to stimulate economic growth and expand local markets. Considerable progress has been made in expanding regional trade through regional bodies such as COMESA, SADC, and the EAC. There are eight other such RECs that have been recognized by the African Union as building blocks for pan-African economic integration. (See Appendix I for details on the Regional Economic Communities [RECs].)

So far regional cooperation in agriculture is in its infancy and major challenges lie ahead. Africa should intensify efforts to use regional bodies as agents of agricultural innovation through measures such as regional specialization. The continent should factitively explore ways to strengthen the role of the RECs in promoting common regulatory standards.

It is not possible to cover the full range of agricultural activities in one volume. This book does not address the important roles that livestock and aquaculture play in Africa. Similarly, it does not deal with innovation in agricultural machinery. But we hope that the systems approach adopted in the book will help leaders and practitioners to anticipate and accommodate other sources of agricultural innovation.[13]

SELECTED ABBREVIATIONS
AND ACRONYMS

AMU	Arab Maghreb Union
AU	African Union
BecANet	Biosciences Eastern and Central Africa Network
CAADP	Comprehensive Africa Agriculture Development Programme
CEN-SAD	Community of Sahel Sahara States
CIMMYT	International Centre for the Improvement of Maize and Wheat
COMESA	Common Market for Eastern and Southern Africa
DFID	Department for International Development
EAC	East African Community
ECCAS	Economic Community of Central African States
ECOWAS	Economic Community of West African States
ICRISAT	International Crops Research Institute for the Semi-Arid Tropics
IGAD	Intergovernmental Authority for Development
IRRI	International Rice Research Institute
NABNet	North Africa Biosciences Network
NCDC	National Cocoa Development Committee

NEPAD	New Partnership for Africa's Development
RECs	Regional Economic Communities
SADC	Southern African Development Community
WABNet	West African Biosciences Network

THE NEW HARVEST

1

THE GROWING ECONOMY

The current global economic crisis, rising food prices, and the threat of climate change have reinforced the urgency to find lasting solutions to Africa's agricultural challenges. Africa is largely an agricultural economy with the majority of the population deriving their income from farming. Agricultural development is therefore intricately linked to overall economic development in African countries. Most policy interventions have focused on "food security," a term that is used to cover key attributes of food such as sufficiency, reliability, quality, safety, timeliness, and other aspects of food necessary for healthy and thriving populations. This chapter outlines the critical linkages between food security, agricultural development, and economic growth and explains why Africa has lagged behind other regions in agricultural productivity. Improving Africa's agricultural performance will require significant political leadership, investment, and deliberate policy efforts.

The Power of Inspirational Leadership

In a prophetic depiction of the power of inspirational models, Mark Twain famously said: "Few things are harder to put up with than the annoyance of a good example." Malawi's remarkable efforts to address the challenges of food security were implemented against the rulebook of economic dogma that preaches against agricultural subsidies to farmers. Malawi's President Bingu wa Mutharika defied these teachings and put in place a series of policy measures that addressed agricultural development and overall economic development. He serves as an example for other African leaders of how aggressive agricultural investment (16% of government spending) can yield increased production and results.

His leadership should be viewed against a long history of neglect of the agricultural sector in Africa. The impact of structural adjustment policies on Malawi's agriculture was evident from the late 1980s.[1] Mounting evidence showed that growth in the smallholder sector had stagnated, with far-reaching implications for rural welfare. The focus of dominant policies was to subsidize consumers in urban areas.[2] This policy approach prevailed in most African countries and was associated with the continued decline of the agricultural sector.

In 2005, over half of the population in Malawi lived on less than a dollar a day, a quarter of the population lacked sufficient food daily, and a third lacked access to clean water. This started to change when Malawi's wa Mutharika took on food insecurity, a dominant theme in the history of the country.[3] His leadership helped to revitalize the agricultural sector and provides an inspiring lesson for other figures in the region who wish to enable and empower their people to meet their most basic needs.

In 2005, Malawi's agricultural sector employed 78% of the labor force, over half of whom operated below subsistence. Maize is Malawi's principal crop and source of nutrition, but for decades, low rainfall, nutrient-depleted soil, inadequate investment, failed privatization policies, and deficient technology led to low productivity and high prices.[4] The 2005 season yielded just over half of the maize required domestically, leaving five million Malawians in need of food aid.

The president declared food insecurity his personal priority and set out to achieve self-sufficiency and reduce poverty, declaring, "Enough is enough. I am not going to go on my knees to beg for food. Let us grow the food ourselves."[5] The president took charge of the ministry of agriculture and nutrition and initiated a systematic analysis of the problem and potential solutions. After a rigorous assessment, the government designed a program to import improved seeds and fertilizer for distribution to farmers at subsidized prices through coupons.

This ambitious program required considerable financial, political, and public support. The president engaged in debate and consultation with Malawi's parliament, private sector, and civil society, while countering criticism from influential institutions.[6] For example, the International Monetary Fund and the United States Agency for International Development (USAID) had fundamentally disagreed with the subsidy approach, claiming that it would distort private sector activities. Other organizations such as the South Africa–based Regional Hunger and Vulnerability Programme questioned the ability of the program to benefit resource-poor farmers.[7] On the other hand the UK Department for International Development (DFID) and the European Union, Norway, Ireland, and later the World Bank, supported the program. Additional support came from China, Egypt, and the Grain Traders and Processors Association. The president leveraged this support and several platforms to explain the program and its intended benefits to the public and their role in the system.[8] With support increasing and the ranks of the hungry swelling, the president devoted approximately US$50 million in discretionary funds and some international sources to forge ahead with the program.[9]

The president's strategy attempted to motivate the particularly poor farmers to make a difference not only for their families, but also for their community and their country. Recognizing the benefits of the program, people formally and informally enforced the coupon system to prevent fraud and corruption. The strategy sought to target smallholder farmers, who face the biggest challenges but whose productivity is essential for improving nutrition and livelihoods.[10]

In 2005–06, the program, coupled with increased rainfall, contributed toward a doubling of maize production, and in 2006–07, the country recorded its highest surplus ever. Prices fell by half, and Malawi began exporting maize to its food insecure neighbors. Learning from experience, the government made a number of adjustments and improvements to the program in its first few years, including stepped-up enforcement of coupon distribution, more effective targeting of subsidies, private sector involvement, training for farmers, irrigation investments, and post-harvest support.

President wa Mutharika's commitment to tackling his nation's most grave problem and development opportunity is a model for channeling power to challenge the *status quo*. Following an integrated approach, the government is devoting 16% of its national budget to agriculture, surpassing the 10% target agreed to in the 2003 Maputo Declaration.[11] His all-too-rare approach to study an issue, develop a solution, and implement it with full force, despite a hostile international environment, demonstrates the difference political will can make. In 2010 he handed over the agriculture portfolio to a line minister.

Linkages Between Agriculture and Economy

Agriculture and economic development are intricately linked. It has been aptly argued that no country has ever sustained rapid economic productivity without first solving the food security challenge.[12] Evidence from industrialized countries as well as countries that are rapidly developing today indicates that agriculture stimulated growth in the nonagricultural sectors and supported overall economic well-being. Economic growth originating in agriculture can significantly contribute to reductions in poverty and hunger. Increasing employment and incomes in agriculture stimulates demand for nonagricultural goods and services, boosting nonfarm rural incomes as well.[13] While future trends in developing countries are likely to be affected by the forces of globalization, the overall thesis holds for much of Africa.

Much of our understanding of the linkages between agriculture and economic development has tended to use a linear approach. Under this model agriculture is seen as a source of input into other sectors of the economy. Resources, skills, and capital are presumed to flow from agriculture to industry. In fact, this model is a central pillar of the "stages of development" that treat agriculture as a transient stage toward industry phases of the economy.[14] This linear view is being replaced by a more sophisticated outlook that recognizes the role of agriculture in fields such as "income growth, food security and poverty alleviation; gender empowerment; and the supply of environmental services."[15] A systems view of economic evolution suggests continuing interactions between agriculture and other sectors of the economy in ways that are mutually reinforcing.[16] Indeed, the relationship between agriculture and economic development is interactive and associated with uncertainties that defy causal correlation.[17]

The Green Revolution continues to be a subject of considerable debate.[18] However, its impact on agricultural productivity and reductions in consumer prices can hardly be disputed. Much of the debate over the impact of the Green Revolution ignores the issue of what would have happened to agriculture in developing countries without it. On the whole, without international research in developing countries, yields in major crops would have been higher in industrialized countries by up to 4.8%. This is mainly because lower production in the developing world would have pushed up prices and given industrialized country farmers incentives to boost their production. It is estimated that crop yields in developing countries would have been up to 23.5% lower without the Green Revolution and that equilibrium prices would have been between 35% and 66% higher in 2000. But in reality prices would have remained constant or risen marginally in the absence of international research. This is mainly because real grain prices actually dropped by 40% from 1965 to 2000.[19]

Higher world prices would have led to the expansion of cultivated areas, with dire environmental impacts. Estimates suggest that crop production would have been up to 6.9% higher in industrialized countries and up to 18.6% lower in developing countries. Over the period, developing countries would have had to increase their food imports by nearly 30% to offset the reductions in production. Without international research, caloric intake in developing countries would have dropped by up to 14.4% and the proportion of malnourished children would have increased by nearly 8%. In other words, the Green Revolution helped to raise the health status of up to 42 million preschool children in developing countries.[20]

It is not a surprise that African countries and the international community continue to seek to emulate the Green Revolution or recommend its variants as a way to address current and future challenges.[21] More important, innovation-driven agricultural growth has pervasive economy-wide benefits as demonstrated through India's Green Revolution. Studies on regional growth linkage have shown strong multiplier effects from agricultural growth to the rural nonfarm economy.[22]

It is for this reason that agricultural stagnation is viewed as a threat to prosperity. Over the last 30 years, agricultural yields and the poverty rate have remained stagnant in sub-Saharan Africa. Prioritizing agricultural development could yield significant, interconnected benefits, particularly in achieving food security and reducing hunger; increasing incomes and reducing poverty; advancing the human development agenda in health and education; and reversing environmental damage.

In sub-Saharan Africa, agriculture directly contributes to 34% of GDP and 64% of employment.[23] Growth in agriculture is at least two to four times more effective in reducing poverty than in other sectors.[24] Growth in agriculture also stimulates productivity in other sectors such as food processing. Agricultural products also compose about 20% of Africa's exports. Given these figures it is no surprise that agricultural research and extension services can yield a 35% rate of return, and irrigation projects a 15%–20% return in sub-Saharan Africa.[25]

Even before the global financial and fuel crises hit, hunger was increasing in Africa. In 1990, over 150 million Africans were hungry; as of 2010, the number had increased to nearly 239 million. Starting in 2004, the proportion of undernourished began increasing, reversing several decades of decline, prompting 100 million people to fall into poverty. One-third of people in sub-Saharan Africa are chronically hungry—many of whom are smallholders. High food prices in local markets price out the poorer consumers—forcing them to purchase less food and less nutritious food, as well as divert spending from education and health and sell their assets. This link of hunger and weak agricultural sector is self-perpetuating. As a World Bank study has shown, caloric availability has a positive impact on agricultural productivity.[26]

Half of African countries with the highest levels of hunger also have among the highest gender gaps. Agricultural productivity in sub-Saharan Africa could increase significantly if such gaps were reduced in school and in the control of agricultural resources such as land. In addition to this critical gender dynamic, the rural-urban divide is also a key component of the agricultural and economic pictures.

Over the last 25 years, growth in agricultural gross domestic product (GDP) in Africa has averaged approximately 3%, but there has been significant variation among countries. Growth per capita, a proxy for farm income, was basically zero in the 1970s and negative from the 1980s into the 1990s. Six countries experienced negative per capita growth. As such, productivity has been basically stagnant over 40 years—despite significant growth in other regions, particularly Asia, thanks to the Green Revolution.[27] Different explanations derive from a lack of political prioritization, underinvestment, and ineffective policies. The financial crisis has exacerbated this underinvestment, as borrowing externally has become more expensive, credit is less accessible, and foreign direct investment has declined.

Only 4% of Africa's crop area is irrigated, compared to 39% in South Asia. Much of rural Africa is without passable roads, translating to high transportation costs and trade barriers. Over 40% of the rural population lives in arid or semi-arid conditions, which have the least agricultural potential. Similarly, about 50 million people in sub-Saharan Africa and 200 million people in North Africa and the Middle East live in areas with absolute water scarcity. Cropland per agricultural population has been decreasing for decades. Soil infertility has occurred due to degradation: nearly 75% of the farmland is affected by excessive extraction of soil nutrients.

One way that farmers try to cope with low soil fertility and yields is to clear other land for cultivation. This practice amounts to deforestation, which accounts for up to 30% of greenhouse gas emissions globally. Another factor leading to increased greenhouse gas emissions is limited access to markets: more than 30% of the rural population in sub-Saharan Africa, the Middle East, and North Africa live more than five hours from a market; another 40% live between two to four hours from a market.

Fertilizer use in Africa is less than 10% of the world average of 100 kilograms per hectare. Just five countries (Ethiopia, Kenya, South Africa, Zimbabwe, and Nigeria) account for about two-thirds of the fertilizer applied in Africa. On the average, sub-Saharan African farmers use 13 kilograms of nutrients per hectare of arable and permanent cropland. The rate in the Middle East and North Africa is 71 kilograms. Part of the reason that fertilizer usage is so low is the high cost of imports and transportation; fertilizer in Africa is two to six times the average world price. This results in low usage of improved seed; as of 2000, about 24% of the cereal-growing area used improved varieties, compared to 85% in East Asia and the Pacific. As of 2005, 70% of wheat crop area and 40% of maize crop area used improved seeds, a significant improvement.

Africa's farm demonstrations show significantly higher average yields compared to national yields and show great potential for improvement in maize. For example, Ethiopia's maize field demonstrations yield over five tons per hectare compared to the national average of two tons per hectare for a country plagued by chronic food insecurity. This potential will only be realized as Africans access existing technologies and improve them to suit local needs.

China's inspirational success in modernizing its agriculture and transforming its rural economy over the last 30 years provided the basis for rapid growth and a substantial improvement in prosperity. From 1978 to 2008 China's economy grew at an annual average rate of about 9%. Its agricultural GDP rose by about 4.6% per year, and farmers' incomes grew by 7% annually. Today, just 200 million small-scale farmers each working an average of 0.6 hectares of land feed a population of 1.3 billion. In the meantime, China was able to limit population growth at 1.07% per year using a variety of government policies. Even more remarkable has been the rate of poverty reduction. China's poverty incidence fell from 31% in 1978 to 9.5% in 1990 and then to 2.5% in 2008. Food security has been dramatically enhanced by the growth and diversification of food production, which outstripped population growth. Agriculture's role in reducing poverty has been three times higher than that of other sectors. Agriculture has therefore been the main force in China's poverty reduction and food security.[28]

Lessons from China show that detailed and sustained focus on small-scale farmers by unleashing their potential and meeting their needs can lead to growth and poverty reduction, even when the basic agricultural conditions are unfavorable. But a combination of clear public policies and institutional reforms are needed for this to happen. The policies and reforms need to be adjusted in light of changing circumstances to bolster the rural economy (through infrastructure services, research support, and farmer education), stimulate off-farm employment, and promote rural-urban migration as rural productivity rises and urban economies expand.

With population in check, China's grain production soon outstripped direct consumption, and policy attention shifted to agricultural diversification and improvement of rural livelihoods. The process was driven by a strong, competent, and well-informed developmental state that could set clear medium and long-term goals and support their implementation.

Despite the historical, geographic, political, social, educational, and cultural differences between China and Africa, there are still many lessons from China's agricultural transformation that can inspire Africa's efforts to turn around decades of low agricultural investment and misguided policies. An African agricultural revolution is within reach, provided the continent can focus on supporting small-scale farmers to help meet national and regional demand for food, rather than rely on expansion of export crops.

While prospects for Africa's global agricultural commodities markets (including cocoa, tea, and coffee) are likely to be brighter than in recent decades, the African food market will grow from US$50 billion in 2010 to US$150 billion by 2030. Currently, food imports are estimated at US$30 billion, up from US$13 billion in the 1990s. Meeting this market with local production will generate the revenue needed to attract additional foreign investment and help in overall economic diversification. Such a transformation will also help expand overall economic development through linkages with urban areas.

China and the OECD's Development Assistance Committee are helping to disseminate lessons from China's experience among African policymakers and practitioners. But they can go further by contributing to the implementation of agricultural strategies developed by African leaders through the Regional Economic Communities (RECs) and other political bodies. At the very least, they should support efforts to strengthen Africa's capacity for evidence-based policy-making and implementation. This will help to create national and regional capacity for strategic thinking and implementation of specific agricultural programs.[29]

The State of African Agriculture

Africa has abundant arable land and labor which, with sound policies, could be translated into increased production, incomes, and food security. This has not materialized because of lack of consistent policies and effective implementation strategies arising from the neglect of the sector. Thus, even though agriculture accounts for 64% of the labor force, over 34% of GDP, and over 20% of businesses in most countries, it continues to be given low priority.[30]

Over the past 40 years, there has been remarkable growth in agricultural production, with per capita world food production growing by 17% and aggregate world food production growing by 145%.[31] However, in Africa, food production is 10% lower today than it was in 1960 because of low levels of investment in the sector. The recent advances in aggregate world productivity have therefore not brought reductions in the incidence of hunger in African countries. Of the 800 million people worldwide lacking adequate access to food, a quarter of them are in sub-Saharan Africa. The number of hungry people has in fact increased by 20% since 1990.

Strategies for transforming African agriculture have to address such challenges as low investment and productivity, poor infrastructure, lack of funding for agricultural research, inadequate use of yield-enhancing technologies, weak linkages between agriculture and other sectors, unfavorable policy and regulatory environments, and climate change.

The path to productivity growth in sub-Saharan Africa will differ considerably from that in irrigated Asian rice and wheat farming systems. Sub-Saharan African agriculture is 96% rain fed and highly vulnerable to weather shocks. And diverse agroecological conditions produce a wide range of farming systems based on many food staples, livestock, and fisheries.

Most agriculture-based countries are small, making it difficult for them to achieve scale economies in research and training. Unless regional markets are better integrated, markets will also be small. Nearly 40% of Africa's population lives in landlocked countries that face transport costs that, on average, are 50% higher than in the typical coastal country.

Vast distances and low population densities in many countries in sub-Saharan Africa make trade, infrastructure, and service provision costly and slow down the emergence of competitive markets. Conversely, areas of low population density with good agricultural potential represent untapped reserves for agricultural expansion.

More than half the world's conflicts in 1999 occurred in sub-Saharan Africa. Although the number of conflicts has declined in recent years, the negative impacts on growth and poverty are still significant. Reduced conflict offers the scope for rapid agricultural growth as demonstrated by Mozambique's recent experience.[32]

The human capital base of the agriculture profession is aging as a result of the decline in support for training during the past 20 years and the HIV/AIDS epidemic. But major accomplishments in rural primary education are ensuring a future generation of literate and numerate African smallholders and nonfarm entrepreneurs. Nevertheless, education has been slow to play a key role in the capacity of farmers to diversify into nonfarm activities.[33]

Despite these common features, the diversity across sub-Saharan African countries and across regions within countries is huge in terms of size, agricultural potential, transport links, reliance on natural resources, and state capacity. The policy agenda will have to be carefully tailored to country-specific circumstances.

Many African governments have treated agriculture as a way of life for farmers who in most cases have no voice in lobbying for an adequate share of public expenditure. Following the Maputo Summit, African countries agreed to devote at least 10% of their public expenditure to agriculture. By 2008 only 19% of African countries had allocated more than 10% of their national expenditure to agricultural development. Many countries hardly reached 4% of GDP and have depended on official development assistance for funding agriculture and other sectors.

Sub-Saharan Africa ranks the lowest in the world in terms of yield-enhancing practices and techniques. Yield-enhancing practices include mechanization, use of agro-chemicals (fertilizers and pesticides), and increased use of irrigated land. The use of these practices and technologies is low in Africa even in comparison to other developing regions. This at least partly explains why crop yields in Africa in general are far below average yields in other parts of the world.

Mechanization is very low, with an average of only 13 tractors per 100 square kilometers of arable land, versus the world average of 200 tractors per 100 square kilometers.[34] In the UK, for example, there are 883 tractors per 1,000 farm workers, whereas in sub-Saharan Africa there are now two per 1,000, which is actually a 50% drop from the 1980 level of three.[35] In Africa, tractor plowing and use of other modern inputs are confined to areas with high market demand or large-scale farms. Therefore, there is considerable variation in the use of these technologies across the continent's RECs. Irrigated land is only 3.6% of total cropland on the continent compared with the world average of 18.4%, while the use of fertilizers is minimal at nine kilograms per hectare compared with the world average of 100 kilograms per hectare. The development of irrigated agriculture is highest in the Common Market for Eastern and Southern Africa (COMESA)—14.4% of arable land—possibly due to the large irrigation projects in Egypt and the Sudan.

Undercapitalization of agriculture as discussed above has given rise to an agricultural sector with a weak knowledge base, resulting in low-input, low-output, and low-value-added agriculture in most cases. Land productivity in Africa is estimated at 42% and 50% of that in Asia and Latin America, respectively. Asia and Latin America have more irrigated land and use more fertilizers and machinery than Africa.

Africa has 733 million hectares of arable land (27.4% of world total) compared with 570 million hectares for Latin America and 628 million hectares for Asia. Only 3.8% of Africa's surface and groundwater is harnessed, while irrigation covers only 7% of cropland (3.6% in sub-Saharan Africa). Clearly, there is considerable scope for both horizontal and vertical expansion in African agriculture.[36]

However, area expansion should not be a priority in view of increased environmental degradation on the continent. Currently, Africa accounts for 27% of the world's land degradation and has 500 million hectares of moderately or severely degraded land. Degradation affects 65% of cropland and 30% of pastureland. Soil degradation is associated with low land productivity. It is mainly caused by loss of vegetation and land exploitation, especially overgrazing and shifting cultivation.

African agriculture is weakly integrated with other sectors such as the manufacturing sector. By promoting greater sectoral linkages, value chain development can greatly enhance job creation, agricultural transformation, and broad-based growth on the continent. Therefore, Africa should take the necessary measures to confront its challenges in this area.

Trends in Agricultural Renewal

Future trends in African agriculture are going to be greatly influenced by developments in the global economy as well as emerging trends in Africa itself. Despite recent upheavals in the global financial system, Africa continues to register remarkable growth prospects. While African economies currently face serious challenges, such as poverty, diseases, and high rates of infant mortality, Africa's collective GDP (at US$1.6 trillion in 2008) is almost equal to that of Brazil or Russia—two emerging markets.[37]

Furthermore, Africa is among the most rapidly growing economic regions in the world today. Its real GDP grew approximately 4.9% per year from 2000 to 2008, and major booming sectors include telecom, banking, and retail, followed by construction and foreign investment. While each African nation faces a unique growth path, a framework for such development has been created by McKinsey Global Institute to address the opportunities and challenges facing various countries in Africa. From 2002 to 2007, the sector share of real GDP growth was spread out. While the resources consist of 24% of the share of GDP, the rest came from other sectors including wholesale (13%), retail trade (13%), agriculture (12%), transportation (10%), and manufacturing (9%).[38]

As agricultural growth has a huge potential for companies across the value chain, overcoming various barriers to raising productivity (such as a lack of advanced seeds, inadequate infrastructure, trade barriers, unclear land rights, lack of technical assistance, and finance for farmers) is a key to increasing the agricultural output from US$280 billion to a projected US$880 billion by 2030.[39]

The picture is therefore promising though uncertain. Several recent successes demonstrate that the link between farm productivity and income growth for the poor indeed operates in Africa. Several countries have exhibited higher agricultural growth rates per capita over the last 10 years. These recent gains in agriculture can be attributed to a better policy environment, increased usage of technology, and higher commodity prices. There are numerous cases that illustrate the ingenious and innovative ways that Africans are overcoming the constraints identified above to strengthen their agricultural productivity and livelihoods.

For example, Ghana has made consistent progress in reducing poverty and hunger. Between 1991–92 and 2006, Ghana nearly halved its poverty rate from over 51% to 28%. Ghana is also the only African country to reduce its Global Hunger Index by more than 50%. The success can be attributed to a better investment climate, policies, and commodity prices. The agricultural sector's rate of growth was higher than both overall GDP and the service sector between 2001 and 2005. Increased land use and productivity among smallholders and cash crop growers in cocoa and horticulture—particularly pineapples—drove growth and welfare improvement.

With success come challenges and lessons learned: inequality has increased, suggesting that the benefits of this growth have not been evenly distributed and that more attention needs to be paid to the rural north. Also, unsustainable environmental degradation and natural resource usage threatens to reverse progress in agriculture and affect other sectors. But the global financial, food, and fuel crises are negatively impacting the agricultural sector and the poor. Prices of inputs and crops have risen by anywhere from 26% to 51% between 2007 and 2008 in real terms. Although cocoa prices are still high for exporters, shea nut prices have fallen—a major source of income for women in the Savannah region.

Social safety net programs (such as cash transfers, school feeding, and national insurance), though, are providing some buffer against the current crises' effects on income and consumption. Ghana's story helps show the importance of locally owned policies and political commitment to sustain agricultural gains and welfare improvement.

Enabling Policy Environment

Although still hindered by unfair international terms of trade, a more favorable policy and macroeconomic environment has helped spur agricultural development in recent years. Countries that have relaxed constraints (such as over-taxation of the agricultural sector) have been able to increase agricultural productivity. For example, a 10% increase in coffee prices in Uganda has helped reduce the number of people living in poverty by 6%.

For less than a few dollars, land-use certificates can be implemented to reduce encroachment and improve soil conservation. For example, Ethiopia's system for community-driven land certification has been one effective way to improve land practices and a potential step toward the much broader reform of land policy that is needed in many African countries.

Here is how it works: communities learn about the certification process and then elect land-use committees. These voluntary committees settle conflicts and designate unassigned plots through a survey, setting up a system for inheritable rights. In a nationwide survey, approximately 80% felt that this certification process effectively fulfilled those tasks as well as encouraged their personal investment in conservation and women's access to resources. The certificates themselves cost US$1 per plot but increase to less than US$3 with mapping and updating using global position system (GPS). Between 2003 and 2005, six million households were issued certificates, demonstrating the scalability.[40] Documenting land rights in this participatory and locally owned way can serve as a model for governments ready to take on meaningful reform.

The initiation of the African-led Comprehensive Africa Agriculture Development Programme (CAADP) by the African Union's NEPAD constitutes a significant demonstration of commitment and leadership. Since 2003, CAADP has been working with the RECs and through national roundtables to promote sharing, learning, and coordination to advance agriculture-led development. CAADP focuses on sustainable land management, rural infrastructure and market access, food supply and hunger, and agricultural research and technology. As of June 2010, CAADP had signed "compacts" with 26 countries. The compacts are products of national roundtables at which priorities are set and roadmaps for implementation are developed. The compacts are signed by all the key partners.

In eastern and southern Africa, COMESA coordinates the CAADP planning and implementation processes at country and regional levels. In doing so, it also collaborates with regional policy networks, such as the Food, Agriculture and Natural Resources Policy Analysis Network) and subregional knowledge systems such as the Regional Strategic Analysis and Knowledge Support Systems, and it utilizes analytical capacity provided through various universities in the region, supported by Michigan State University.

In close coordination with national CAADP processes, a regional CAADP compact is being developed. Its aim is to design a Regional Investment Program on Agriculture that will focus on developing key regional value chains and integrating value chain development into corridor development programs. At a national level, the priority programs developed include those in the area of research and dissemination of productivity-enhancing technologies to promote knowledge-based agricultural practices applying the innovation systems approach to develop and strengthen linkages between generators, users, and intermediaries of technological knowledge.

Regional Imperatives

The facilitation of regional cooperation is emerging as a basis for diversifying economic activities in general, and leveraging international partnerships in particular. Many of Africa's individual states are no longer viable economic entities; their future lies in creating trading partnerships with neighboring countries.

Many African countries are either relatively small or land-locked, thereby lacking the financial resources needed to invest in major infrastructure projects. Their future economic prospects depend on being part of larger regional markets. Increased regional trade in agricultural products can help them stimulate rural development and enhance their technological competence through specialization. Existing RECs offer them the opportunity to benefit from rationalized agricultural activities. They can also benefit from increased harmonization of regional standards and sanitary measures.[41]

African countries have adopted numerous regional cooperation and integration arrangements, many of which are purely ornamental. The roles of bigger markets in stimulating techno-logical innovation, fostering economies of scale arising from infrastructure investments, and the diffusion of technical skills into the wider economy are some of the key gains Africa hopes to derive from economic integration. In effect, science and innovation are central elements of the integration agenda and should be made more explicit.

The continent has more than 20 regional agreements that seek to promote cooperation and economic integration at subregional and continental levels. They range from limited cooperation among neighboring states in narrow political and economic areas to the ambitious creation of an African common market. They focus on improving efficiency, expanding the regional market, and supporting the continent's integration into the global economy. Many of them are motivated by factors such as the small size of the national economy, a land-locked position, and poor infrastructure. Of all Africa's regional agreements, the African Union (AU) formally recognizes eight RECs. These RECs represent a new economic governance system for Africa and should be strengthened.

The Common Market for Eastern and Southern Africa, in particular, illustrates the importance of regional integration in Africa's economic development and food security. The 19-member free trade area was launched in 2000 and at 420 million people accounts for nearly half of Africa's population. It has a combined GDP of US$450 billion and is the largest and most vibrant free trade area in Africa, with intra-COMESA trade estimated at US$14.3 billion in 2008. COMESA aims to improve economic integration and business growth by stan-dardizing customs procedures, reducing tariffs, encouraging investments, and improving infrastructure. COMESA launched its customs union on June 7, 2009, in Victoria Falls Town and has initiated work on a Common Investment Area to facilitate cross-border and foreign direct investment. COMESA plans to launch its common market in 2015, and in this regard it already has a program for liberalization of trade in services. The program prioritizes liberalization of infrastructure services, namely, communication, transport, and financial services. Other subsectors will be progressively liberalized.

The strength of the RECs lies in their diversity. Their objec-tives range from cooperation among neighboring states in narrow political and economic areas to the ambitious creation of political federations. Many of them are motivated by factors such as the small size of the national economy, a landlocked position, or poor infrastructure. Those working on security, for example, can learn from the Economic Community of West African States (ECOWAS) which has extensive experience dealing with conflict in Ivory Coast, Liberia, and Sierra Leone.[42]

Other RECs have more ambitious plans. The EAC, for example, has developed a road map that includes the use of a common currency and creation of single federal state. In July 2010 the EAC launched its Common Market by breaking barriers and allowing the free movement of goods, labor, services, and capital among its member states. The EAC Common Market has a combined GDP of US$73 billion. Through a process that began with the establishment of the EAC Customs Union, the Common Market is the second step in a four-phase roadmap to make the EAC the strongest economic, social, cultural, and political partnership in Africa. EAC's economic influence extends to neighboring countries such as Sudan, Democratic Republic of the Congo, and Somalia. The Common Market will eliminate all tariff and nontariff barriers in the region and set up a common external tax code on foreign goods. It will also enhance macroeconomic policy coordination and harmonization as well as the standardization of trade practices. It is estimated that East Africa's GDP is will grow 6.4% in 2011, making it the fastest growing region in Africa.[43]

The region has already identified agriculture as one of its strategic areas. In 2006 the EAC developed an Agriculture and Rural Development Policy that provides a framework for improving rural life over the next 25 years by increasing productivity output of food and raw materials, improving food security, and providing an enabling environment for regional and international trade. It also covers the provision of social services such as education, health and water, development of support infrastructure, power, and communications. The overall vision of the EAC is to attain a "well developed agricultural sector for sustainable economic growth and equitable development."[44] Its mission is to "support, promote and facilitate the development, production and marketing of agricultural produce and products to ensure food security, poverty eradication and sustainable economic development."[45] Such institutions, though nascent, represent major innovations in Africa's economic and political governance and deserve the fullest support of the international community.

Conclusion

This chapter has examined the critical linkages between agriculture and economic development in Africa. It opened with a discussion of the importance of inspirational leadership in effecting change. This is particularly important because much of the large body of scientific and technical knowledge needed to promote agricultural innovation in Africa is available. It is widely acknowledged that institutions play an important role in shaping the pace and direction of technological innovation in particular and economic development in general. Much has been written on the need to ensure that the right democratic institutions are in place as prerequisites for agricultural growth.[46] But emerging evidence supports the importance of entrepreneurial leadership in promoting agricultural innovation as a matter of urgency and not waiting until the requisite institutions are in place.[47] This view reinforces the important role that entrepreneurial leadership plays in fostering the co-evolution between technology and institutions.

Fundamentally, "it would seem that one can understand the role of institutions and institutional change in economic growth only if one comes to see how these variables are connected to technological change."[48] This is not to argue that institutions and policies do not matter. To the contrary, they do and should be the focus of leadership. What is important is that the focus should be on innovation. The essence of entrepreneurial leadership of the kind that President wa Mutharika has shown in Malawi points to the urgency of viewing institutions and economic growth as interactive and co-evolutionary. The rest of this book will examine these issues in detail.

2

ADVANCES IN SCIENCE, TECHNOLOGY, AND ENGINEERING

The Green Revolution played a critical role in helping to over-come chronic food shortages in Latin America and Asia. The Green Revolution was largely a result of the creation of new institutional arrangements aimed at using existing technology to improve agricultural productivity. African countries are faced with enormous technological challenges. But they also have access to a much larger pool of scientific and technical knowledge than was available when the Green Revolution was launched in the 1950s. The aim of this chapter is to review major advances in science, technology, and engineering and identify their potential for use in African agriculture. This exploration will also include an examination of local innovations as well as indigenous knowledge. It will cover fields such as information and communications technology, genetics, ecology, and geographical sciences. It will emphasize the convergence of these and other fields and their implications for African agriculture.

Innovation and Latecomer Advantages

African countries can utilize the large aggregation of knowl-edge and know-how that has been amassed globally in their efforts to improve their access to and use of the most cutting-edge technology. While Africa is currently lagging in the utilization and accumulation of technology, its countries have the ability not only to catch up to industrial leaders but also to attain their own level of research growth.

Advocates of scientific and technical research in developing countries have found champions in the innovation platforms of nanotechnology, biotechnology, information and communication technology (ICT), and geographic information systems (GIS). Through these four platform technologies, Africa has the opportunity to promote its agenda concurrent with advances made in the industrialized world. This opportunity is superior to the traditional catching-up model, which has led to slower development and kept African countries from reaching their full potential. These technologies are able to enhance technological advances and scientific research while expanding storage, collection, and transmission of global knowledge. This chapter explores the potential of ICT, GIS, nanotechnology, and biotechnology in Africa's agricultural sector and provides examples of where these platform technologies have already created an impact.

Contemporary history informs us that the main explanation for the success of the industrialized countries lies in their ability to learn how to improve performance in a variety of fields—including institutional development, technological adaptation, trade, organization, and the use of natural resources. In other words, the key to success is putting a premium on learning and improving problem-solving skills.[1]

Every generation receives a legacy of knowledge from its predecessors that it can harness for its own advantage. One of the most critical aspects of a learner's strategy is using that legacy. Each new generation blends the new and the old and thereby charts its own development path within a broad technological trajectory, making debates about potential conflicts between innovation and tradition irrelevant.[2]

At least three key factors contributed to the rapid economic transformation of emerging economies. First, these countries invested heavily in basic infrastructure, including roads, schools, water, sanitation, irrigation, clinics, telecommunications, and energy.[3] The investments served as a foundation for technological learning. Second, they nurtured the development of small and medium-sized enterprises (SMEs).[4] Building these enterprises requires developing local operational, repair, and maintenance expertise, and a pool of local technicians. Third, government supported, funded, and nurtured higher education institutions as well as academies of engineering and technological sciences, professional engineering and technological associations, and industrial and trade associations.[5]

The emphasis on knowledge should be guided by the view that economic transformation is a process of continuous improvement of productive activities, advanced through business enterprises. In other words, government policy should focus on continuous improvement aimed at enhancing performance, starting with critical fields such as agriculture for local consumption and extending to international trade.

This improvement indicates a society's capacity to adapt to change through learning. It is through continuous improvement that nations transform their economies and achieve higher levels of performance. Using this framework, with government functioning as a facilitator for social learning, business enterprises will become the locus of learning, and knowledge will be the currency of change.[6] Most African countries already have in place the key institutional components needed to make the transition to being a player in the knowledge economy. The emphasis should therefore be on realigning the existing structures and creating new ones where they do not exist.

The challenge is in building the international partnerships needed to align government policy with the long-term technological needs of Africa. The promotion of science and technology as a way to meet human welfare needs must, however, take into account the additional need to protect Africa's environment for present and future generations.

The concept of "sustainable development" has been advanced specifically to ensure the integration of social, economic, and environmental factors in development strategies and associated knowledge systems.[7] Mapping out strategic options for Africa's economic renewal will therefore need to be undertaken in the context of sustainable development strategies and action plans.

There is widespread awareness of rapid scientific advancement and the availability of scientific and technical knowledge worldwide. This growth feeds on previous advances and inner self-propelling momentum. In fact, the spread of scientific knowledge in society is eroding traditional boundaries between scientists and the general public. The exponential growth in knowledge is also making it possible to find low-cost, high-technology solutions to persistent problems.

Life sciences are not the only areas where research could contribute to development. Two additional areas warrant attention. The continent's economic future crucially depends on the fate and state of its infrastructure, whose development will depend on the contributions of engineering, materials, and related sciences. It is notable that these fields are particularly underdeveloped in Africa and hence could benefit from specific missions that seek to use local material in activities such as road construction and maintenance. Other critical pieces involve expanding the energy base through alternative energy development programs. This sector is particularly important because of Africa's past investments, its available human resources, and its potential to stimulate complementary industries that provide parts and services to the expansion of the sector. Exploiting these opportunities requires supporting policies.

Advances in science and technology will therefore make it possible for humanity to solve problems that have previously been in the realms of imagination. This is not a deterministic view of society but an observation of the growth of global knowledge and the feasibility of new technical combinations that are elicited by social consciousness. This view would lead to the conclusion that Africa has the potential to access more scientific and technical knowledge than the more advanced countries had in their early stages of industrialization.

Recent evidence shows the role that investment in research plays in Africa's agricultural productivity. For example, over the past three decades Africa's agricultural productivity grew at a much higher rate (1.8%) than previously calculated, and technical progress, not efficiency change, was the primary driver of this crucial growth.[8] This finding reconfirms the critical role of research and development (R&D) in agricultural productivity. The analysis also lends further support for the key role pro-trade reforms play in determining agricultural growth.

Within North Africa, which has experienced the highest of the continent's average agricultural productivity growth (of 3.6% per year), Egypt stands out as a technology leader, as the gross majority of its agricultural growth has been attributable to technical investments and progress, not efficiency gains. A similar trend stressing the importance of technical progress and R&D has been seen in an additional 20 African countries that have experienced annual productivity growth rates over 2%.

This evidence shows that the adaptive nature of African agricultural R&D creates shorter gestation lags for the payout from R&D when compared to basic research. This makes the case for further investment even more important.[9] On the whole, African agricultural productivity increased on net by 1.4% in the 1970s, 1.7% in the 1980s, and 2.1% in the 1990s. While growth in these decades can be attributed largely to major R&D investments made in the 1970s, declines in productivity growth in the 2000s are attributed to decreased R&D investments in the late 1980s and 1990s. With an average rate of return of 33% for 1970–2004, sustained investment in adaptive R&D is demonstrated to be a crucial element of agricultural productivity and growth.

Evidence from other regions of the world tells the same story in regard to specific crops. For example, improvements in China's rice production illustrate the significant role that technical innovation plays in agricultural productivity. Nearly 40% of the growth in rice production in13 of China's rice growing provinces over the 1978–84 period can be accounted for by technology adoption. Institutional reform could explain 35% of the growth. Nearly all the growth in the subsequent 1984–90 period came from technology adoption. These findings suggest that the impact of institutional reform, though significant, has previously been overstated.[10] The introduction of new agricultural technologies went hand in hand with institutional reform.

Chinese agriculture has grown rapidly during the past several decades, with most major crops experiencing increased yield, area harvested, and production. Between 1961 and 2004, maize, cotton, wheat, and oilseed production had an average growth rate of 4% per annum, while rice production growth was 2.8% per annum. However, the growth rates for wheat crops in terms of area harvested, yield, and production were less than 1% per annum between 1961 and 2004. A decomposition of the sources of wheat production growth indicates that growth between 1961 and 2004 was primarily driven by yield growth, with modest contribution from crop area. In the case of North Korea, achieving self-sufficiency in food production has been one of the most important objectives of its economic strategy. Even so, the period between 1961 and 2004 saw minimal growth in area harvested for rice and soybeans and negative growth for maize and wheat.[11]

Since the early 1960s, South Korea has transformed itself from a low-income agrarian economy into a middle-income industrialized "miracle," and the agricultural sector in South Korea has not been immune to the tremendous structural change. The decline in production in South Korea has continually been driven by area contraction, whereas increased production was driven primarily by yield change in some cases and by area change in other instances. In an environment of poor natural resources and subsequent encroachment of the industrial and services sector on agriculture, Taiwan has experienced negative growth rates of rice, wheat, and oilseed area harvested from 1961 until 2004. In most cases, production growths were due to increased yields and production decreases were due to contractions in crop area.

Generic or Platform Technologies

Information and Communications Technologies

While information and communication technologies in industrialized countries are well developed and historically established, ICTs in developing countries have traditionally been "based on indigenous forms of storytelling, song and theater, the print media and radio."[12] Despite Africa's current deficiency in more modern modes of communication and information sharing, the countries benefit greatly from the model of existing technologies and infrastructure.

In addition to the specific uses that will be explored in the rest of this section, ICTs have the extremely significant benefit of providing the means for developing countries to contribute to and benefit from the wealth of knowledge and research available in, for example, online databases and forums. The benefits of improved information and communication technologies range from enhancing the exchange of inter- and intra-continental collaborations to providing agricultural applications through the mapping of different layers of local landscapes.

Mobile Technology

Sub-Saharan Africa has 10 times as many mobile phones as landlines in operation, providing reception to over 60% of the population. Much of this growth in cell phone use—as much as 49% annually from 2002 through 2007—coincided with economic growth in the region. It is estimated that every 10% growth in mobile phones can raise up to 1.5% in GDP growth. There are five ways in which mobile phone access boosts microeconomic performance: reducing search costs and therefore improving overall market efficiency, improving productive efficiency of firms, creating new jobs in telecommunications-based industries, increasing social networking capacity, and allowing for mobile development projects to enter the market.[13]

Mobile phones cut out the opportunity costs, replacing several hours of travel with a two-minute phone call and also allow firms and producers to get up-to-date information on demand. They also redistribute the economic gains and losses per transaction between consumers and producers. This reduction in the costs of information gathering creates an ambiguous net welfare gain for consumers, producers, and firms. Similarly, mobile phones make it easier for social networks to absorb economic shocks. Family and kinship relationships have always played an important role in African society, and mobile phones strengthen this already available "social infrastructure," allowing faster communication about natural disasters, epidemics, and social or political conflicts.

The use of more mobile phones creates a demand for additional employment. For instance, formal employment in the private transport and communications sector of Kenya rose by 130% between 2003 and 2007, as mobile phone use rose about 49% annually during that period. While there is a measurable growth in formal jobs, such as hotline operators who deliver information on agricultural techniques, there is also growth in the informal sector, including the sale of phone credit and "pay-as-you-go" phones, repair and replacement of mobile phone hardware, and operation of phone rental services in rural areas. New employment opportunities also come through the mobile development industry.

Africa's advantage over countries like the United States in avoiding unnecessary infrastructure costs is especially exemplified in the prevalence of mobile technologies, which have replaced outdated landline connectivity. Mobile phones have a proven record in contributing to development, as illustrated by the associated rise in the rate of mobile phone use, averaging 65% annually over the last five years.[14] Because mobile phones are easy to use and can be shared, this mobility has revolutionized and facilitated processes like banking and disease surveillance.

The potential and current uses of mobile technology in the agricultural sector are substantial and varied. For instance, local farmers often lacked the means to access information regarding weather and market prices making their job more difficult and decreasing their productivity. With cellular phones comes cheap and convenient access to information such as the cost of agricultural inputs and the market prices for crops.

The desire for such information has led to the demand for useful and convenient mobile phone-based services and applications: "New services such as *AppLab*, run by the Grameen Foundation in partnership with Google and the provider MTN Uganda, are allowing farmers to get tailored, speedy answers to their questions. The initiative includes platforms such as *Farmer's Friend*, a searchable database of agricultural information, *Google SMS*, a question and answer texting service and *Google Trader*, a SMS-based 'marketplace' application that helps buyers and sellers find each other."[15] Applications such as these coupled with the increased usage of cellular phones have reduced the inefficiencies and unnecessary expenses of travel and transportation.

Simple services like text messaging have likewise led to an expansion of access and availability of knowledge. This service has been enhanced by nonprofit Kiwanja.Net's development of a software package allowing the use of text message services where there is no Internet, so long as one has a computer. Other advantages include the ability to set up automatic replies to messages using keywords (for example, in the case of a scheduled vaccination). Nongovernmental organization (NGO) managers, doctors, and researchers around the world have enthusiastically picked up this technology and used it to solve their communication challenges—from election monitoring, to communicating health and agricultural updates, to conducting surveys, to fund-raising—the list is endless.

From opening access to research institutes to facilitating business transactions, few technologies have the potential to revolutionize the African agricultural sector as much as the Internet. The demand for Internet service in Africa has been shown in the large increases in Internet usage (over 1,000% between 2000 and 2008) and through the fiber-optic cable installed in 2009 along the African east coast by Seacom. In 2010 MaIN OnE launched a new cable to serve the west coast of Africa. Just as with mobile phones, the Internet will have a transformative impact on the operations of businesses, governments, NGOs, farmers, and communities alike.

Until recently Africa was served by an undersea fiber-optic cable only on the west coast and in South Africa. The rest of the continent relied on satellite communication. The first undersea fiber-optic cable, installed by Seacom, reached the east African coast in July 2009. The US$600 million project will reduce business costs, create an e-commerce sector, and open up the region to foreign direct investment.

New industries that create content and software are likely to emerge. This will in turn stimulate demand for access devices. A decade ago it cost more than US$5,000 to install one km of standard fiber-optic cable. The price has dropped to less than

US$300. However, for Africa to take advantage of the infrastructure, the cost of bandwidth must decline. Already, Internet service providers are offering more bandwidth for the same cost. For example, in 2009 MTN Business in South Africa cut the cost of bandwidth by up to 50%.

Access to broadband is challenging Africa's youth to demonstrate their creativity and the leaders to provide a vision of the role of infrastructure in economic transformation. The emergence of Safaricom's M-PESA service—a revolutionary way to transmit money by mobile phones—is an indicator of great prospects for using new technologies for economic improvement.[16] In fact, these technologies are creating radically new industries such as branchless banks that are revolutionizing the service sector.[17]

The diffusion of GIS is creating new opportunities for development in general and Africa in particular, with regard to agriculture.[18] An example of the potential GIS has for agriculture is the digitalization of more than 20 million land records under the Bhoomi Project in India's State of Karnataka, which led to improved and more available information on land rights and land use innovation. But the implications of the Bhoomi Project did not just stop here. Because of the availability of such geospatial information, bankers became more inclined to provide crop loans and land disputes began declining, allowing farmers to invest in their land without fear. The success of this project has inspired the government of India to establish the National Land Records Modernisation Programme to do the same for the entire country.[19] Not only does this show the applicability and usefulness of information technologies in agriculture, but it also provides an option to be considered by African countries.

Unlike biotechnology and nanotechnology, ICT and GIS do not have as much risk of being overregulated or reviled for being a great unknown. However, regulation of ICT and GIS will be necessary—and keeping clients and users secure will be a challenge as more and more of Africa becomes connected to the international network. Balancing privacy with the benefits of sharing knowledge will probably be one of the largest challenges for these sectors, especially between countries and companies.

When introducing new technologies in the field, policy makers should consider ones that are low cost and easily accessible to the farmers who will use them. They should ideally capitalize on techniques that farmers already practice and involve support for scaling up and out, rather than pushing for expensive and unfamiliar practices.

New technologies should also allow farmers to be flexible according to their own capacities, situations, and needs. They should require small initial investments and let farmers experiment with the techniques to decide their relative success. If farmers can achieve success with a small investment early on, they are likely to devote more resources to the technique later as they become committed to the practice. An example of this type of technological investment is the planting pits that increased crop production and fostered more productive soil for future years of planting.

Mobile technologies are poised to influence a wide range of sectors. For example, by 2011 the One Laptop per Child (OLPC) Foundation had provided about two million rugged, low-cost, low-power XO laptops to school children around the world. The laptops come with software and content designed to foster collaborative and interactive child-centered learning. Advances in complementary technologies and increased availability of digital books will make education more mobile. Learning will no longer be restricted to classrooms. Emerging trends in mobile technology will also transform heath care. Portable ultrasound devices produced by firms such as General Electric, SonoSite, and Masimo will help to reduce the cost of health care. Technological convergence will also simplify the use of new technologies and make them more widely accessible.

Biotechnology

Biotechnology—technology applied to biological systems—has the promise of leading to increased food security and sustainable forestry practices, as well as improving health in developing countries by enhancing food nutrition. In agriculture, biotechnology has enabled the genetic alteration of crops, improved soil productivity, and enhanced natural weed and pest control. Unfortunately, such potential has largely remained untapped by African countries.

In addition to increased crop productivity, biotechnology has the potential to create more nutritious crops. About 250 million children suffer from vitamin A deficiency, which weakens their immune systems and is the biggest contributor to blindness among children.[20] Other vitamins, minerals, and amino acids are necessary to maintain healthy bodies, and a deficiency will lead to infections, complications during pregnancy and childbirth, and impaired child development. Biotechnology has the potential to improve the nutritional value of crops, leading both to lower health care costs and higher economic performance (due to improved worker health).

Tissue culture has not only helped produce new rice varieties in Africa but has also helped East Africa produce pest- and disease-free bananas at a high rate. The method's ability to rapidly clone plants with desirable qualities that are disease-free is an exciting prospect for current and future research on improved plant nutrition and quantities. Tissue culture has also proved to be useful in developing vaccines for livestock diseases, especially the bovine disease rinderpest. Other uses in drug development are currently being explored.

Tissue culture of bananas has had a great impact on the economy of East African countries since the mid-1990s. Because of its susceptibility to disease, banana has always been a double-edged sword for the African economies like that of

Uganda, which consumes a per capita average of one kilogram per day. For example, when the Black Sigatoka fungus arrived in East Africa in the 1970s, banana productivity decreased as much as 40%. Tissue culture experimentation allowed for quick generation of healthy plants and was met with great success. Since 1995, Kenyan banana production has more than doubled, from 400,000 to over one million tons in 2004, with average yield increasing from 10 tons per hectare to 30–50 tons.

Marker-assisted selection helps identify plant genome sections linked to genes that affect desirable traits, which allows for the quicker formation of new varieties. This technique has been used not only to introduce high-quality protein genes in maize but also to breed drought-tolerant plant varieties. An example of a different application of this method has been the development of maize resistant to Maize streak virus. While the disease has created a loss of 5.5 million tons per year in maize production, genetic resistance is known and has the potential of greatly raising production. The uptake of genetically modified (GM) crops is the fastest adoption rate of any crop technology, increasing from 1.7 million hectares in 1996 to 134 million hectares in 2009, and an 80-fold increase over the period.[21]

Recent increases among early adopting countries have come mainly from the use of "stacked traits" (instead of single traits in one variety or hybrid). In 2009, for example, 85% of the 35.2 million hectares of maize grown in the United States was genetically modified, and three-quarters of this involved hybrids with double or triple stacked traits. Nearly 90% of the cotton growth in the United States, Australia, and South Africa is GM and, of that, 75% has double-stacked traits.

In 2009, there were 14 million farmers growing GM crops in 25 countries around the world, of whom over 90% were small and resource-poor farmers from developing countries. Most of the benefits to such farmers have come from cotton. For example, over the 2002–08 period, *Bacillus thuringiensis* (Bt) cotton added US$5.1 billion worth of value to Indian farmers, cut insecticide use by half, helped to double yield and turned the country from a cotton importer into a major exporter.[22]

Africa is steadily joining the biotechnology revolution. South Africa's GM crop production in corn stood at 2.1 million hectares in 2009, an increase of 18% over the previous year. Burkina Faso grew 115,000 hectares of *Bt* cotton the same year, up from 8,500 in 2008. This was the fastest adoption rate of a GM crop in the world that year. In 2009, Egypt planted nearly 1,000 hectares of *Bt* maize, an increase of 15% over 2008.[23]

African countries, by virtue of being latecomers, have had the advantage of using second-generation GM seed. Monsanto's Genuity™ Bollgard II® (second generation) cotton contains two genes that work against leaf-eating species such as armyworms, budworms, bollworms, and loopers. They also protect against cotton leaf perforators and saltmarsh caterpillars. Akin to the case of mobile phones, African farmers can take advantage of technological leapfrogging to reap high returns from transgenic crops while reducing the use of chemicals. In 2010 Kenya and Tanzania announced plans to start growing GM cotton in view of the anticipated benefits of second-generation GM cotton. The door is now open for revolutionary adoption of biotechnology that will extend to other crops as technological familiarity and economic benefits spread.

There is also a rise in the adoption of GM crops in Europe. In 2009, six European countries (Spain, Czech Republic, Portugal, Romania, Poland, and Slovakia) planted commercial *Bt* maize. Trends in Europe suggest that future decisions on GM crops will be driven by local needs as more traits become available. For example, crops that tolerate various stresses such as drought are likely to attract interest among farmers in Africa. The Water Efficient Maize for Africa project, coordinated by the African Agricultural Technology Foundation in collaboration with the International Centre for the Improvement of Maize and Wheat (CIMMYT) and Monsanto and support from the Howard Buffett Foundation and the Bill and Melinda Gates Foundation, is an example of such an initiative that also brings together private and public actors.[24]

This case also represents new efforts by leading global research firms to address the concerns of resource-poor farmers, a subtheme in the larger concern over the contributions of low-income consumers.[25] Other traits that improve the efficiency of nitrogen uptake by crops will also be of great interest to resource-poor farmers. Other areas that will attract interest in developing new GM crops will include the recruitment of more tree crops into agriculture and the need to turn some of the current grains into perennials.[26]

Trends in regulatory approvals are a good indicator of the future of GM crops. By 2009, some 25 countries had planted commercial GM crops and another 32 had approved GM crop imports for food and feed use and for release into the environment. A total of 762 approvals had been granted for 155 events (unique DNA recombinations in one plant cell used to produce entire GM plants) for 24 crops. GM crops are accepted for import in 57 countries (including Japan, the United States, Canada, South Korea, Mexico, Australia, the Philippines, the European Union (EU), New Zealand, and China). The majority of the events approved are in maize (49) followed by cotton (29), canola (15), potato (10), and soybean (9).[27]

Because of pest attacks, cotton was, until the early 1990s, the target of 25% of worldwide insecticide use.[28] Recombinant DNA engineering of a bacterial gene that codes for a toxin lethal to bollworms resulted in pest-resistant cotton, increasing profit and yield while reducing pesticide and management costs.[29] Countries like China took an early lead in adopting the technology and have continued to benefit from reduced use of pesticides.[30]

While GM crops have the potential to greatly increase crop and livestock productivity and nutrition, a popular backlash against GM foods has created a stringent political atmosphere under which tight regulations are being developed. Much of the inspiration for restrictive regulation comes from the Cartagena Protocol on Biosafety under the United Nations Convention on Biological Diversity.[31] The central doctrine of the Cartagena Protocol is the "precautionary principle" that empowers governments to restrict the release of products into the environment or their consumption even if there is no scientific evidence that they are harmful.

These approaches differ from food safety practices adopted by the World Trade Organization (WTO) that allow government to restrict products when there is sufficient scientific evidence of harm.[32] Public perceptions are enough to trigger a ban on such products. Those seeking stringent regulation have cited uncertainties such as horizontal transfer of genes from GM crops to their wild relatives. Others have expressed concern that the development of resistance to herbicides in GM crops results in "super-weeds" that cannot be exterminated using known methods. Some have raised fears about the safety of GM foods to human health. Other concerns include the fear that farmers would be dependent on foreign firms for the supply of seed.[33]

The cost of implementing these regulations could be beyond the reach of most African countries.[34] Such regulations have extended to African countries, and this tends to conflict with the great need for increased food production. As rich countries withdraw funding for their own investments in agriculture, international assistance earmarked for agricultural science has diminished.[35]

In June 1999, five European Union members (Denmark, Greece, France, Italy, and Luxembourg) formally declared their intent to suspend authorization of GM products until rules for labeling and traceability were in place. This decision followed a series of food-related incidents such as "mad cow disease" in the UK and dioxin contamination in Belgium. These events undermined confidence in regulatory systems in Europe and raised concerns in other countries. Previous food safety incidents tended to shape public perceptions over new scares.[36] In essence, public responses reactions to the GM foods were shaped by psychological factors.[37] Much of this was happening in the early phases of economic globalization when risks and benefits were uncertain and open to question, including the very moral foundations of economic systems.[38]

Much of this debate occurred at a time of increased awareness about environmental issues and there had been considerable investment in public environmental advocacy to prepare for the 1992 United Nations Conference on Environment and Development in Rio de Janeiro. These groups teamed up with other groups working on issues such as consumer protection, corporate dominance, conservation of traditional farming practices, illegal dumping of hazardous waste, and promotion of organic farming to oppose the introduction of GM crops. The confluence of forces made the opposition to GM crops a global political challenge, which made it easier to try to seek solutions through multilateral diplomatic circles.

The moratorium was followed by two important diplomatic developments. First, the EU used its influence to persuade its trading partners to adopt similar regulatory procedures that embodied the precautionary principle. Second, the United States, Canada, and Argentina took the matter to the WTO for settlement in 2003.[39] Under the circumstances, African countries opted for a more restrictive approach partly because they had stronger trade relations with the EU and were therefore subject to diplomatic pressure. Their links with the United States were largely through food aid programs.[40]

In 2006, the WTO issued its final report on the dispute; the findings were largely on procedural issues and did not resolve the root cause of the debate such as the role of the "precautionary principle" in WTO law and whether GM foods were substantially equivalent to their traditional counterparts.[41] But by then a strong anti-biotechnology culture had entrenched itself in most African countries.[42] For example, even after developing a GM potato resistant to insect damage, Egypt refused to approve it for commercial use. This resistance grew to the point that Africa ceased to accept unmilled GM maize from the United States as food aid. A severe drought in 2001–02 left 15 million Africans with severe food shortages; countries like Zimbabwe and Zambia turned down shipments of GM maize, fearing that the kernels would be planted instead of eaten. Unlike the situation in rich countries, GM foods in developing countries have the potential to revolutionize the lots of suppliers and consumers. In order to take full advantage of the many potentials of biotechnology in agriculture, Africa should consider whether aversion to and overregulation of GM production are warranted.[43]

In Nigeria, the findings of a study on biotechnology awareness demonstrate that while respondents have some awareness of biotechnology techniques, this is not the case for biotechnology products.[44] Most of the respondents are favorably disposed to the introduction of GM crops and would eat GM foods if they are proven to be significantly more nutritious than non-GM foods. The risk perception of the respondents suggests that although more people are in favor of the introduction of GM crops, they, however, do not consider the current state of Nigeria's institutional preparedness satisfactory for the approval and release of genetically modified organisms (GMOs).

However, it is important to consider that African farmers will not grow successful crops if prices are low or dropping. Additionally, complications with regulation and approval of GM crops make obtaining commercial licenses to grow certain crops difficult. Also, neighboring countries must often approve similar legislation to cover liabilities that might arise from cross-pollination by wind-blown pollen, for example. Biosafety regulations often stall developments in the research of GM crops and could have negative impacts on regional trade.[45]

For these reasons, approval and use of potentially beneficial crops are often difficult. However, despite potential setbacks, biotechnology has the potential to provide both great profits and the means to provide more food to those who need it in Africa. Leaders in the food industry in parts of Africa prefer to consider the matter on a case-by-case basis rather adopt a generic approach to biosafety.[46] In fact, the tendency in regulation of biotechnology appears to follow more divergence paths reflecting unique national and regional attributes.[47] This is partly because regulatory practices and trends in biotechnology development tend to co-evolve as countries seek a balance between the need to protect the environment and human safety and fostering technological advancement.[48]

Advancements in science have allowed scientists to insert characteristics of other plants into food crops. Since the introduction of GM crops in 1996, over 80%–90% of soybean, corn, and cotton grown in the United States today comes from GM crops. Despite their widespread use, there are limited data on their environmental, economic, and social impact.[49]

Herbicide-resistance GM crops have fewer adverse effects on the environment than natural crops, but often at the cost of farming efficiency. The growth of most crops requires the use of toxic chemical herbicides, but GM crops utilize an organic compound called glyphosphate to combat weeds. While less dangerous toxins are entering the environment, weeds are developing a resistance to glyphosphate in soybean, corn, and cotton crops, reducing farming efficiency and raising prices on these goods.

GM corn and cotton have helped reduce the amount of insecticides entering the environment. Insecticides are harmful to most insects, regardless of their impact, positive or negative, on crops. Genetically engineered corn and cotton produce *Bacillus thuringiensis* (Bt) toxins, which kill the larvae of beetles, moths, and flies. New genetic hybrids are introduced frequently to reduce the threat of a *Bt* resistant pest. Since 1996, insecticide use has decreased while *Bt* corn use has grown considerably. While the environmental benefits are clear, GM crops pose a threat to farmers who rely on nonengineered crops. Interbreeding between crops is difficult to stop, so regulatory agencies must set clear standards on how much GM material is allowed to be present in organic crops.

The rapid adoption of GM crops seems to indicate that they offer great economic benefits for farmers. In general, farmers experience lower production costs and higher yields because weed control is cheaper and fewer losses are sustained from pests. GM crops are safer to handle than traditional chemical pesticides and herbicides, increasing worker safety and limiting the amount of time workers spend in the field. While the supply-side benefits for farmers are clear, it is not completely understood how genetic modification affects the market value for these crops. Holding technological achievement constant, any gains tend to dissipate over time.

The United States has benefited by being among the first adopters of GM crops. In a similar vein, it is not clear what economic effects planting GM crops will have on farmers who do not adopt the technology. Livestock farmers are one of the largest customers of corn and soybean for feed and should receive the largest benefits of the downward pressure on prices from transgenic crops, yet no study has been conducted on such effects. Similarly, it is possible that the growing use of GM crops leaves many pests resistant to chemicals to ravage the fields of nonadopters, forcing them to use higher concentrations of dangerous chemicals or more expensive forms of control. In the future, new public policy will be needed to develop cost-effective methods of controlling the growing weed resistance to glyphosphate.

It is important to recognize that developing countries face a separate set of risks from those of industrialized countries.

For example, new medicines could have different kinds and levels of effectiveness when exposed simultaneously to other diseases and treatments. Similarly, "new technologies may require training or monitoring capacity which may not be locally available, and this could increase risks associated with the technology's use."[50] This has been demonstrated where a lack of training in pesticide use has led to food contamination, poisoning, and pesticide resistance. In addition, the lack of consistent regulation, product registration, and effective evaluation are important factors that developing Africa will need to consider as it continues its exploration of these platform technologies. Probably the most significant research and educational opportunities for African countries in biotechnology lie in the potential to join the genomics revolution as the costs of sequencing genomes drop. When James Watson, co-discover of the DNA double-helix, had his genome sequenced in 2008 by 454 Life Sciences, the price tag was US$1.5 million. A year later a California-based firm, Applied Biosystems, revealed that it has sequenced the genome of a Nigerian man for under US$60,000. In 2010 another California-based firm, Illumina, announced that it had reduced the cost to about US$20,000.

Dozens of genomes of agricultural, medical, and environmental importance to Africa have already been sequenced. These include human, rice, corn, mosquito, chicken, cattle, and dozens of plant, animal, and human pathogens. The challenge facing Africa is building capacity in bioinformatics to understanding the location and functions of genes. It is through the annotation of genomes that scientists can understand the role of genes and their potential contributions to agriculture, medicine, environmental management, and other fields. Bioinformatics could do for Africa what computer software did for India. The field would also give African science a new purpose and help to integrate the region into the global knowledge ecology. This opportunity offers Africa another opportunity for technological leapfrogging.

Nanotechnology

Of the four platform technologies discussed in this chapter, nanotechnologies are the least explored and most uncertain. Nanotechnology involves the manipulation of materials and devices on a scale measured by the billionths of a meter. The results of research in nanotechnology have produced substances of both unique properties and the ability to be targeted and controlled at a level unseen previously. Thus far, applications to agriculture have largely been theoretical, but practical projects have already been explored by both the private and public sectors in developed, emerging, and developing countries.[51]

For example, research has been done on chemicals that could target one diseased plant in a whole crop. Nanotechnology has the potential to revolutionize agriculture with new tools such as the molecular treatment of diseases, rapid disease detection, and enhancement of the ability of plants to absorb nutrients. Smart sensors and new delivery systems will help to combat viruses and other crop pathogens. Increased pesticide and herbicide effectiveness as well as the creation of filters for pollution create more environmentally friendly agriculture process. While countries like the United States and China have been at the forefront of nanoscience research expenditure and publications, emerging countries have engaged in research on many of the applications, from water purification to disease diagnosis.

Water purification through nanomembranes, nanosensors, and magnetic nanoparticles have great, though currently cost prohibitive, potential in development, particularly in countries like Rwanda, where contaminated water is the leading cause of death. However, the low energy cost and high specificity of filtration has lead to a push for research in water filtration and purification systems like the Seldon WaterStick and WaterBox. Developed by U.S.-based Seldon Laboratories, these products require low energy usage to filter various pathogens and chemical contaminants and are already in use by aid workers in Rwanda and Uganda.[52]

One of the most promising applications of nanotechnology is low-cost, energy-efficient water purification. Nearly 300 million people in Africa lack access to clean water. Water purification technologies using reverse osmosis are not available in much of Africa, partly because of high energy costs. Through the use of a "smart plastic" membrane, the U.S.-based Dias Analytic Corporation has developed a water purification system that could significantly increase access to clean water and help to realize the recent proclamation by the United Nations that water and sanitation are fundamental human rights.

The capital costs of the NanoClear technology are about half the cost of using reverse osmosis water purification system. The new system uses about 30% less energy and does not involve toxic elements. The system is modular and can be readily scaled up on demand. A first-generation pilot plant opened in Tampa, Florida, in 2010 to be followed later in the year by the deployment of the first fully operational NanoClear water treatment facility in northern China. The example of NanoClear illustrates how nanotechnology can help provide clean water, reduces energy usage, and charts an affordable course toward achieving sustainable development goals.[53]

The potential for technologies as convenient as these would revolutionize the lifestyles of farmers and agricultural workers in Africa. Both humans and livestock would benefit from disease-free, contaminant-free water for consumption and agricultural use.

The cost-prohibitive and time-intensive process of diagnosing disease promises to be improved by nanotechnological disease diagnostics. While many researchers have focused on human disease diagnosis, developed and developing countries alike have placed an emphasis on livestock and plant pathogen identification in the interest of promoting the food and agricultural industry. Nanoscience has offered the potential of convenient and inexpensive diagnosis of diseases that would otherwise take time and travel. In addition, the convenient nanochips would be able to quickly and specifically identify pathogens with minimal false diagnoses. An example of this efficiency is found in the EU funded Optolab Card project, whose kits allows a reduction in diagnosis time from 6–48 hours to just 15 minutes.

Technology Monitoring, Prospecting, and Research

Much of the debate on the place of Africa in the global knowledge economy has tended to focus on identifying barriers to accessing new technologies. The basic premise has been that industrialized countries continue to limit the ability of developing countries to acquire new technologies by introducing restrictive intellectual property rights. These views were formulated at a time when technology transfer channels were tightly controlled by technology suppliers, and developing countries had limited opportunities to identify the full range of options available to them. In addition, they had limited capacity to monitor trends in emerging technologies. But more critically, the focus on new technologies as opposed to useful knowledge hindered the ability of developing countries to create institutions that focus on harnessing existing knowledge and putting it to economic use.

In fact, the Green Revolution and the creation of a network of research institutes under the Consultative Group on International Agricultural Research (CGIAR) represented an important example of technology prospecting. Most of the traits used in the early breeding programs for rice and wheat were available but needed to be adapted to local conditions. This led to the creation of pioneering institutions such as CIMMYT in Mexico and the International Rice Research Institute (IRRI) in the Philippines.[54]

Other countries have used different approaches to monitor, identify, and harness existing technologies, with a focus on putting them to commercial use. One such example is the Fundación Chile, established in 1974 by the country's Minister for Economic Cooperation, engineer Raúl Sáez. The Fundación Chile was set up as joint effort between the government and the International Telephone and Telegraph Corporation to promote research and technology acquisition. The focus of the institution was to identify existing technologies and match them to emerging business opportunities. It addressed a larger goal of helping to diversity the Chilean economy and created new enterprises based on imported technologies.[55]

But unlike their predecessors who had to manage technological scarcity, Africa's challenge is managing an abundance of scientific and technological knowledge. The rise of the open access movement and the growing connectivity provided by broadband Internet now allows Africa to dramatically lower the cost of technology searches. But such opportunities require different technology acquisition strategies. First, they require the capacity to assess the available knowledge before it becomes obsolete. Second, such assessments have to take into account the growing convergence of science and technology.[56] There is also an increasing convergence between different disciplines.

Moreover, technology assessments must now take into account social impacts, a process that demands greater use of the diverse disciplines.[57] Given the high rate of uncertainty associated with the broader impact of technology on environment, it has become necessary to incorporate democratic practices such as public participation in technology assessments.[58] Such practices allow the public to make necessary input into the design of projects. In addition, they help to ensure that the risks and benefits of new technologies are widely shared.

Reliance on imported technology is only part of the strategy. African countries are just starting to explore ways to increase support for domestic research. This theme should be at the center of Africa's international cooperation efforts.[59] These measures are an essential aspect of building up local capacity to utilize imported technology. This insight is important because the capacity to harness imported technology depends very much on the existence of prior competence in certain fields. Such competence may lie in national research institutes, universities, or enterprises. The pace of technology absorption is likely to remain low in countries that are not making deliberate efforts to build up local research capacity, especially in the engineering sciences. One way to address this challenge is to start establishing regional research funds that focus on specific technology missions.

Conclusion

The opportunities presented by technological abundance and diversity as well as greater international connectivity will require Africa to think differently about technology acquisition. It is evident that harnessing existing technologies requires a more detailed understanding of the convergence between science and technology as well the various disciplines. In addition, it demands closer cooperation between the government, academia, the private sector, and civil society in an interactive process. Such cooperation will need to take into account the opportunities provided by the emergence of Regional Economic Communities as building blocks for Africa's economic integration. All the RECs seek to promote various aspects of science, technology and innovation in general and agriculture in particular. In effect, it requires that policy makers as well as practitioners think of economies as innovation systems that evolve over time and adapt to change. The next chapter will elaborate the idea of innovation systems and their implications for agricultural development and regional integration in Africa.

3

AGRICULTURAL INNOVATION SYSTEMS

The use of emerging technology and indigenous knowledge to promote sustainable agriculture will require adjustments in existing institutions.[1] New approaches will need to be adopted to promote close interactions between government, business, farmers, academia, and civil society. The aim of this chapter is to identify novel agricultural innovation systems of relevance to Africa. It will examine the connections between agricultural innovation and wider economic policies. Agriculture is inherently a place-based activity and so the chapter will outline strategies that reflect local needs and characteristics. Positioning sustainable agriculture as a knowledge-intensive sector will require fundamental reforms in existing learning institutions, especially universities and research institutes. Most specifically, key functions such as research, teaching, extension, and commercialization need to be much more closely integrated.

The Concept of Innovation Systems

Agriculture is considered central to African economies, but it is treated like other sectors, each with their own distinctive institutions and with little regard for their relationship with the rest of the economy.[2] This view is reinforced by traditional approaches, which argue that economic transition occurs in stages that involve the transfer of capital from the agricultural to the industrial sector. Both the sector and stage approaches conceal important linkages between agriculture and other sectors of the economy.

A more realistic view is to treat economies as "systems of innovation." The process of technological innovation involves interactions among a wide range of actors in society, who form a system of mutually reinforcing learning activities. These interactions and the associated components constitute dynamic "innovation systems."[3] Innovation systems can be understood by determining what within the institutional mixture is local and what is external. Open systems are needed, in which new actors and institutions are constantly being created, changed, and adapted to suit the dynamics of scientific and technological creation.[4] The concept of a system offers a suitable framework for conveying the notion of parts, their interconnectedness, and their interaction, evolution over time, and emergence of novel structures. Within countries the innovation system can vary across localities. Local variations in innovation levels, technology adoption and diffusion, and the institutional mix are significant features of all countries.

An innovation system can be defined as a network of organizations, enterprises, and individuals focused on bringing new products, new processes, and new forms of organization into economic use, together with the institutions and policies that affect their behavior and performance. The innovation systems concept embraces not only the science suppliers but the totality and interaction of actors involved in innovation. It extends beyond the creation of knowledge to encompass the factors affecting demand for and use of knowledge in novel and useful ways.[5]

Government, the private sector, universities, and research institutions are important parts of a larger system of knowledge and interactions that allows diverse actors with varied strengths to come together to pursue broad common goals in agricultural innovation. In many African countries, the state still plays a key role in directing productive activities. But the private sector is an increasingly important player in adapting existing knowledge and applying it to new areas.

The innovation systems concept is derived from direct observations of countries and sectors with strong records of innovation. It has been applied to agriculture in developing countries only recently, but it appears to offer exciting opportunities for understanding how a country's agricultural sector can make better use of new knowledge and for designing alternative interventions that go beyond research system investments.[6]

Systems-based approaches to innovation are not new in the agricultural development literature. The study of technological change in agriculture has always been concerned with systems, as illustrated by applications of the national agricultural research system (NARS) and the agricultural knowledge and information system (AKIS) approaches. However, the innovation systems literature is a major departure from the traditional studies of technological change that are often used in NARS- and AKIS-driven research.[7]

The NARS and AKIS approaches, for example, emphasize the role of public sector research, extension, and educational organizations in generating and disseminating new technologies. Interventions based on these approaches traditionally focused on investing in public organizations to improve the supply of new technologies. A shortcoming of this approach is that the main restriction on the use of technical information is not just supply or availability but also the limited ability of innovative agents to absorb it. Even though technical information may be freely accessible, innovating agents have to invest heavily to develop the ability to use the information.

While both the NARS and AKIS frameworks made critical contributions to the study of technological change in agriculture, they are now challenged by the changing and increasingly globalized context in which sub-Saharan African agriculture is evolving. There is need for a more flexible framework for studying innovation processes in developing-country agriculture—a framework that highlights the complex relationships between old and new actors, the nature of organizational learning processes, and the socioeconomic institutions that influence these relationships and processes.

The agricultural innovation system maps out the key actors and their interactions that enable farmers to obtain access to technologies. The "farm firm" is at the center of the agricultural innovation system framework, and the farmer as the innovator could be made less vulnerable to poverty when the system enables him to access returns from his innovative efforts. The agricultural innovation system framework presents a demand-driven approach to agricultural R&D. This transcends the perception of the role of public research institutions as technology producers and farmers as passive users by viewing the public laboratory-farmer relationships as an interactive process governed by several institutional players that determine the generation and use of agricultural innovation. There is opportunity for a participatory and multi-stakeholders approach to identifying issues for agricultural R&D, and agricultural technology could thus be developed with active farmers' participation and understanding of the application of new technologies. The agricultural innovation system approach as an institutional framework can be fostered depending on the institutional circumstances and historical background of the national agricultural development strategies.[8]

This brings us to the agricultural innovation system (AIS) framework. The AIS framework makes use of individual and collective absorptive capabilities to translate information and knowledge into a useful social or economic activity in agriculture. The framework requires an understanding of how individual and collective capabilities are strengthened, and how these capabilities are applied to agriculture. This suggests the need to focus far less on the supply of information and more on systemic practices and behaviors that affect organizational learning and change. The approach essentially unpacks systemic structures into processes as a means of strengthening their development and evolution.

Recent discussions of innovation capacity have argued that capacity development in many countries involves two sorts of tasks. The first is to create networks of scientific actors around research themes such as biotechnology and networks of rural actors around development themes such as dryland agriculture. The second is to build links between these networks so that research can be used in rural innovation. A tantalizing possibility is that interventions that unite research-led and community-based capacity could cost relatively little, add value to existing investments, meet the needs of the poor, and achieve very high returns.

Innovation systems in action

University-Industry Linkages

Trends in university-industry linkages (UILs) in Nigeria illustrate three ways in which university-industry collaboration has been experienced in the Nigerian agro-food processing sector. They are principal agent demand-driven, multi-stakeholder problem based, and arms-length consultancy. The examples of university-industry interactions in these three modes are regarded as glimpses of hope demonstrating that universities and firms in Nigeria can be made to work together to build capacity for innovation. However, while the first two modes have contributed to innovative outcomes involving the diffusion and commercialization of local R&D results, the third mode has not engendered innovation.[9]

The first mode of UIL identified as "principal agent demand-driven" is the UNAAB-Nestlé Soyabean Popularization and Production Project which has been a case of interaction between the University of Agriculture Abeokuta (UNAAB) and Nestlé Nigeria since 1999. In this case Nestlé employed UNAAB to help address its challenges in demand for soybeans. Due to its research and extension activities, UNAAB presumably has a knowledge advantage over Nestlé in the area of local sourcing of soybeans. Nestlé Nigeria employs about 1,800 people and soybeans are one of its major raw materials used especially for baby foods. The firm has been the only major external donor and industrial partner with UNAAB. It is thus plausible to consider the principal agent in this case of UIL as Nestlé, and the driver of the UIL as demand for soybeans.

The main objective of the UIL is to stimulate sustainable interest of farmers in soybean production with a view to increasing their capacity to produce seeds of industrial quality and consequently to improving their socioeconomic status. The three specific objectives of the project include ensuring that the soybean becomes acceptable and properly integrated into the existing farming systems in the southwestern part of Nigeria; promoting massive production of high quality grains that would meet the needs and quality standards required by Nestlé Nigeria on a continuous basis; and improving the welfare of the farmers through the income that could be generated from soybean production.

The university-industrial linkage can be initially traced to an R&D partnership under a tripartite agreement for soybean breeding between UNAAB, the International Institute of Tropical Agriculture (IITA), Ibadan, and Nestlé Nigeria in the early 1990s. Nestlé Nigeria financed the soybean breeding project. The aim of the project was to obtain soybeans of high quality that fit Nestlé Nigeria's requirements and also to produce significantly improved yields. The research team achieved this objective with the breeding of Soya 1448–2E. This initial partnership ended in 1996.

Around 1999, Nestlé Nigeria came back to ask UNAAB if there could be ways of further partnership. UNAAB told Nestlé that the previous research collaboration had established that soybeans can also be grown in southwest Nigeria. UNAAB thus started a project with Nestlé Nigeria on popularization of soybeans in southwest Nigeria. Nestlé Nigeria had previously believed that soybeans can be grown only in northern Nigeria as it was thought that rain was damaging to soybeans just before harvest. Though this is generally true, UNAAB had demonstrated that soybeans could be profitably harvested in spite of the rains in the southwest.

There are a number of benefits for such university-industry linkages. Learning by interaction between UNAAB scientists and Nestlé Nigeria farm managers and farmers contributed significantly to building capacity for innovation especially at the farm level. It produced improved quality seeds and grains and a new process for growing soybeans. Nestlé Nigeria saved costs by finding alternative to the inefficient Nestlé Nigeria farms located in northern Nigeria and secured a regular supply of high-quality soybeans from farmers in the UIL. The system boosted UNAAB's extension activities resulting in the popularization of its model of soybean cultivation in southwest Nigeria, which in turn became an important soybean producing region. Overall, the linkages improved the livelihoods of the people in the region and enhanced technology adoption for soybean processing, especially threshing technology.

The second mode of UIL identified as "multi-stakeholder problem based" is the Cassava Flash Dryer Project. The project involved one large privately owned integrated farm (Godilogo Farm, Ltd.) that had an extensive cassava plantation and a cassava processing factory; three universities including the University of Agriculture, Abeokuta, the University of Ibadan, and the University of Port Harcourt; the IITA; and the Raw Material Research and Development Council (RMRDC).

Cassava is Africa's second most important food staple, after maize, in terms of calories consumed, and it is widely acknowledged as a crop that holds great promise for addressing the challenges of food security and welfare improvement. Nigeria is currently the world largest producer of cassava. The Presidential Initiative on Cassava Production and Export (PICPE) was officially launched in 2004. Under PICPE the government promotes the diversification of the economy through industrial processing of cassava to add value and achieve significant export of cassava products. Support for research on cassava processing and cassava products was a major aspect of PICPE. Through IITA, PICPE brought together cassava stakeholders to address the challenge of cassava production and industrial processing, which included the design and fabrication of a cassava flash dryer.

Though the principle of flash drying is well known in engineering theory and practice, the principle has so far not been applied in the design of engineering equipment used in processing indigenous agricultural crops in Nigeria. This design gap is perhaps because the engineering properties of most of the Nigerian crops are yet to be determined. The flash dryers available in the market are designed for agricultural products that are grown in industrialized countries that manufacture flash dryers. For example, flash dryers commonly used in Nigeria were originally designed for drying Irish potatoes or maize. They are usually modified with the help of foreign technical partners for use in cassava processing. Attempts to adapt foreign flash dryers have resulted in considerably low performance and frequent equipment breakdowns. This was the experience of Godilogo Farms, Ltd. that had used a flash dryer imported from Brazil, because the design was unable to handle the drying of cassava to required moisture or water content. It was not originally designed to handle cassava but temperate root crops such as Irish potatoes. Efforts by a Brazilian engineer invited from abroad to adapt the flash dryer to effectively process cassava failed.

The farm's cassava plantation could supply its cassava processing factory 250 days of cassava inputs. The farm also has an engineering workshop or factory for equipment maintenance and components fabrication.

The main objective of the cassava flash dryer project was to design and fabricate an efficient cassava flash dryer that can withstand the stress of the local operating environment. The frustration of Godilogo Farm with its imported flash dryer motivated the farm's management to support the cassava flash dryer project. After the farm management was convinced that the flash dryer research team constituted under the PICPE-IITA cassava processing research project had a feasible design, Godilogo Farm made available its engineering facilities and funds for building a cassava flash dryer in situ at the farm's cassava processing factory.

The new locally designed and fabricated cassava flash dryer can produce 250 kilograms of cassava flour per hour. The RMRDC funded the official commissioning of the new flash dryer at Godilogo Farms, Obudu, Cross Rivers State, on August 19, 2008. IITA and PICPE provided the initial funding under the IITA Integrated Cassava Project; the Root and Tuber Extension Program supported the design team's visit to collect data from existing flash drying centers; Godilogo paid for the fabrication of the plant and part sponsorship of the researchers' living costs; and RMRDC provided logistical support for several trips by the design team including sponsorship of the commissioning.

The main outcome of the UIL is the celebrated local design and fabrication of the first medium-sized cassava flash dryer in Nigeria. The technological learning generated was unprecedented in local fabrication of agro-food processing equipment, and there is evident improvement in capacity for innovation in agro-food processing. In the course of the project, there was interactive learning through experimentation by the research team. The impact of government policy through PICPE and government support for the project through RMRDC demonstrated the crucial role of government as a mediator or catalyst for UIL and innovation. Knowledge flows and user feedbacks also played important roles in the success of the university-industry linkage.

Wider Institutional Linkages

Understanding the network relationships and institutional mechanisms that affect the generation and use of innovation in the traditional sector is critical for enhancing the welfare of the poor and overall economic development. Nigeria's development policy emphasizes making agriculture and industrial production the engine of growth. In recent years the revitalization of the cocoa industry through the cocoa rebirth initiative launched in February 2005 has been a major focus of government.[10]

The program essentially aimed at generating awareness of the wealth creation potentials of cocoa, promoting increases in production and industrial processing, attracting youth into cocoa cultivation, and helping to raise funds for the development of the industry. By applying the analytical framework of the agricultural system of innovation it is easier to trace the process of value-addition in the cocoa agro-industrial system, examining the impact of the cocoa rebirth initiative and identifying the actors critical for strengthening the cocoa innovation system in Nigeria.

Cocoa production is a major agricultural activity in Nigeria; and R&D aimed at improving cocoa production and value-addition has long existed at the Cocoa Research Institute of Nigeria (CRIN) and notable faculties of agriculture in Nigerian universities and colleges of agriculture. However, while the export of raw cocoa beans has continued to thrive, innovation in cocoa production and the industrial processing of cocoa into intermediate and consumer products have been limited.

The cocoa innovation system in Nigeria is still relatively weak. There is a role for policy intervention in stimulating interaction among critical agents in this agricultural innovation system. In particular, linkages and interactions between four critical actors (individual cocoa farmers, cocoa processing firms, CRIN, and the National Cocoa Development Committee) in the cocoa rebirth program were identified as being responsible for the widespread adoption of CRIN's newly developed genetically improved cocoa seedlings capable of a yield exceeding 1.8 tonnes per hectare per year. This is in stark contrast to previous experiences of CRIN, which has been unable to commercialize many of its research findings. Periodic joint review of the activities of each of these actors and active participation in specific projects that are of common interest may further innovation especially in value-addition to cocoa beans.

The adoption and diffusion of improved cocoa seedlings under the cocoa rebirth initiative thrives on subsidies provided by government. While subsidies for agricultural production in a developing country such as Nigeria may not be discouraged, it is important to have a phased program of subsidy withdrawal on the cocoa seedlings program when it is certain that farmers have proven the viability and economic importance of the new variety. This should result in a market-driven diffusion that will be healthy for the sustainable growth of the cocoa industry.

Despite success with the diffusion of cocoa seedlings, the findings show that although export is a major concern of the cocoa processing firms, and this appeared to have led to close interactions of the firms with the National Export Promotion Council (NEPC), the export strategy has not been effectively linked with the cocoa rebirth initiative. In order to further encourage export by the cocoa processing firms, it would be good to integrate the NEPC export incentives into the cocoa rebirth initiative within the cocoa innovation system framework. Moreover, the NEPC should also adopt an innovation system approach to export strategy. This would essentially begin by emphasizing demonstrable innovative activities of firms as an important requirement for the firms to benefit from export incentives.

The involvement of the financial sector in the cocoa innovation system is identified as a main challenge. Though the financial sector is aware of the significance of innovation for a competitive economy, its response to the cocoa rebirth initiative has been slow due to perceived relatively low return on investments. It is suggested that the publicly owned Bank of Industry and the Central Bank of Nigeria (CBN) should provide leadership in investing in innovative new start-ups in cocoa processing and in carefully identified innovative ideas or projects in existing cocoa processing firms. This demonstration should be carried out in partnership with interested commercial banks with the CBN guaranteeing the banks' investment in the project. Once the banks are convinced that innovative initiatives in firms are able to provide satisfactory returns on investment, they should be open to investing in such projects.

Skills deficiency is a major constraint on the cocoa innovation system. The result suggests that skills development in the areas of cocoa farm management and the operation of modern cocoa processing machinery would be particularly useful in enhancing cocoa output and the performance of cocoa processing firms. In this respect, renewed efforts are needed by the educational and training institutions to improve on the quality and quantity of skills being produced for cocoa processing firms.

As part of the cocoa rebirth initiative, special training programs should be organized for skills upgrading and new skills development relevant to the cocoa industry. Another important constraint on the cocoa innovation system arises from the difficulty in implementing the demand-side aspects of the cocoa rebirth initiative, such as serving free cocoa beverages in primary schools and using cocoa-based beverages in government offices, practices that should stimulate innovative approaches to increased local processing of cocoa and manufacture of cocoa-based products.

Clusters as Local Innovation Systems

Theory, evidence, and practice confirm that clusters are important source of innovation.[11] Africa is placing considerable emphasis on the life sciences. There is growing evidence that innovation in the life sciences has a propensity to cluster around key institutions such as universities, hospitals, and venture capital firms.[12] This logic could be extended to thinking about other opportunities for clustering which include agricultural regions. Essentially, clusters are geographic concentrations of interconnected companies and institutions in a particular field. Clusters encompass an array of linked industries and other entities important to competition. They include, for example, suppliers of specialized inputs such as components, machinery, and services, and providers of specialized infrastructure.

Often clusters extend downstream to channels and customers and laterally to manufacturers of complementary products, as well as to companies in related industries either by skills, technologies, or common inputs. Finally, many clusters include governmental and other institutions—such as universities, standard-setting agencies, think tanks, vocational training providers, and trade associations—that provide specialized training, education, information, research, and technical support.[13] The co-evolution of all actors supports the development of dynamic innovation systems, which accelerate and increase the efficiency of knowledge transfer into products, services, and processes and promote growth. As clusters enable the flow of knowledge and information between enterprises and institutions through networking they form a dynamic self-teaching system and they speed up innovation. Local knowledge develops that responds to local needs—something that rivals find hard to imitate.

Although much of the recent literature on clusters focuses on small- to medium-sized high-tech enterprises in advanced industrial countries, a smaller school of literature has already begun expanding the study of clusters to include agricultural innovation. Clusters can and often do emerge anywhere that the correct resources and services exist. However, central to the idea of clusters is that positive "knowledge spillovers" are more likely to occur between groups and individuals that share spatial proximity, language, culture, and other key factors usually tied to geography.

Contrary to scholars who argue that the Internet and other information technologies have erased most barriers to knowledge transfer, proponents of cluster theory argue that geography continues to dominate knowledge development and transfer, and that governments seeking to spark innovation in key sectors (including agriculture) should therefore consider how to encourage the formation and growth of relevant clusters. A key intuition in this argument is that informal social interactions and institutions play a central role in building trust and interpersonal relationships, which in turn increase the speed and frequency of knowledge, resource, and other input sharing.

In developing countries, clusters are present in a wide range of sectors and their growth experiences vary widely, from being stagnant and lacking competitiveness to being dynamic and competitive. This supports the view that the presence of a cluster does not automatically lead to positive external effects. There is therefore a need to look beyond the simple explanation of proximity and cultural factors, and to ask why some clusters prosper and what specifically explains their success.

Shouguang Vegetable Cluster, China

China has a long history of economic clusters in sectors as diverse as silk, porcelain, high technology, and agriculture.[14] One of China's most successful agricultural clusters is the vegetable cluster in Shandong Province. This "Vegetable City," is a leading vegetable production, trading, and export center. Its 53,000-hectare vegetation plantation produces about four million tons annually. Shouguang was one of the poorest areas in the Shandong province until the early 1980s, when vegetable production started. Today five state- and provincial-level agricultural demonstration gardens and 21 nonpolluted vegetable facilities have been established. More than 700 new vegetable varieties have been introduced from over 30 countries and regions. Shouguang also hosts China's largest vegetable seed facility aimed at developing new variety. The facility is co-sponsored by the China Agricultural University. Over the years, vegetable production increased, leading to the emergence of an agro-industrial cluster that has helped to raise per capita income for Shouguang's previously impoverished rural poor.[15] The cluster evolved through four distinctive phases.

In the first emergence phase (1978 to 1984) Shouguang authorities launched programs for massive vegetable planting as a priority for the local development agenda. Shouguang had three main advantages that helped it to emerge as a leading vegetable cluster. These included a long history and tradition of vegetable production, rising domestic and international demand for vegetables, and higher profits exceeding revenue from crops such as rice and wheat. In 1983 Shouguang's vegetable production exceeded 450 tonnes. The local market could not absorb it all, so about 50 tonnes went to waste. The loss prompted Shouguang to construct a vegetable market the following year, thereby laying the foundation for the next phase.

In the second phase of the development of the cluster, local government officials used their authority to bring more peasants and clients into the new market. For example, the officials persuaded the Shengli Oil Field, China's second largest oil base, to become a customer of Shouguang vegetables. This procurement arrangement contributed to the market's early growth. The authorities also helped to set up more than 10 small agricultural product markets around the central wholesale market, creating a market network in the city. The markets directly benefited thousands of local farmers. Despite these developments, high demand for fresh vegetables in winter exceeded the supply.

The third phase of the development of the cluster was associated with rapid technological improvements in greenhouses and increased production. In 1989, Wang Leyi, chief of a village in Shouguang, developed a vegetable greenhouse for planting in the winter, characterized by low cost, low pollution, and high productivity. This inspired peasants to adopt the technology and led to incremental improvements in the construction and maintenance of greenhouses. Communication among peasants and the presence of local innovators helped to spread the new technology. By the end of 1996, Shouguang had 210,000 greenhouses, and the vegetable yield had grown to 2.3 million tons. The Shouguang government focused on promoting food markets. It helped to create more than 30 large specialized markets and 40 large food-processing enterprises. In 1995 the central government authorized the creation of the "Green Channel," an arrangement for transporting vegetables from Shouguang to the capital, Beijing. The transportation and marketing network evolved to include the "Green Channel," the "Blue Channel" (ocean shipping), the "Sky Channel" (air transportation), and the "Internet Channel."

After 1997 the cluster entered its fourth development phase, which involved the establishment of international brand names. The internationalization was prompted by the saturation of domestic markets and rising nontariff trade barriers such as strict and rigorous standards. International safety standards and consumer interest in "green products" prompted Shouguang to establish 21 pollution-free production bases. Foreign firms such as the Swiss-based Syngenta Corporation played a key role in upgrading planting technologies, providing new seed and offering training to peasants. This was done through the Shouguang Syngenta Seeds Company, a joint venture between Syngenta and the local government. Syngenta signed an agreement with the Ministry of Agriculture's National Agricultural Technical Extension and Service Center to train farmers in modern techniques. Since 2000, the one-month (starting April 20) Shouguang vegetable fair has encapsulated and perpetuated this cluster's many successes and has become one of China's premier science and technology events.

Rice Cluster in Benin

Entrepreneurship can spur innovations, steer innovation processes, and compel the creation of an innovation-enabling environment while giving rise to and sustaining the innovation system. Entrepreneurial venture is an embedded power that steers institutions, stimulates learning, and creates or strengthens linkages that constitute the pillars of innovation systems. The dissemination of New Rices for Africa (NERICA) in Benin illustrates what can be considered a "self-organizing innovation system."[16] Through the unique approach combining the innovation systems approach and entrepreneurship theory, this section describes the process by which a class of entrepreneurs took the lead in the innovation process while creating the basis for a NERICA-based system of innovation to emerge.

Benin, which is located in West Africa, covers an area of 112,622 square kilometers and has nearly 8.2 million inhabitants. Its landscape consists mostly of flat to undulating plains but also includes some hills and low mountains. Agriculture is the predominant basis of the country's weak economy; although only contributing 32% of the GDP (as compared to 53.5% of the service sector and 13.7% of the industrial sector), it employs about 65% of the active population.

Despite relatively favorable production environments, Benin's domestic production is weak and meets only 10%–15% of the country's demand for rice. Different people attribute this to different causes, such as policies and institutions that are not suited to supporting domestic production against importations or low quality of products. Irrigation possibilities are not fully exploited, despite the fact that rice production is traditionally rain-fed. There is also minimum input, with improper seeding and lack of fertilizers, pesticides, and herbicides.

NERICA is the brand name of a family of improved rice varieties specially adapted to the agroecological conditions of Africa. It is a hybrid that combines the best traits of two rice species: the African *Oryza glaberrima* and the Asian *Oryza sativa*. It has certain advantages over other species such as high yields, quick maturity, and resistance to local biotic and abiotic stresses such as droughts and iron toxicity. It also has 25% higher protein content than international standard varieties. And it is more responsive to fertilizers. Due to these advantages, different groups that wanted to change the status quo of Benin's agriculture sought to introduce NERICA. They included the government of Benin, the Banque Régionale de Solidarité (BRS), agro-industrial firms such as Tunde Group and BSS-Société Industrielle pour la Production du Riz (BSS-SIPRi), as well as nongovernmental entities such as Songhaï, Projet d'Appui au Développement Rural de l'Ouémé (PADRO), and Vredeseilanden (VECO). These organizations worked closely together to bring to the task skills, knowledge, and interests that could not be found in one entity.

A simple introduction of all of these organizations helps to clarify how they converged on NERICA in their pursuit of agricultural innovation. Songhaï is a socioeconomic and rural development NGO specializing in agricultural production, training, and research. It supports an integrated production system that promotes minimal inputs and the use of local resources. Songhaï was one of the first pioneers of NERICA production in Benin, largely because it was challenged to endorse a framework conducive to rice production as a profitable commodity.

Songhaï came in contact with BRS as it was seeking to fund skilled, competent, and innovative economic agents with sound business plans. Songhaï fit the bill perfectly. Tunde Group was NERICA's production hub and BSS-SIPRi is an enterprise specializing in NERICA seed and paddy production. PADRO and VECO are NGOs from France and Belgium, respectively. PADRO worked with the extension agency, farmer organizations, and micro-finance establishments, and indirectly with the Ministry of Agriculture. VECO focused on culture, communication, sustainable agriculture, and food security.

All of these separate organizations came together through NERICA to challenge Benin's agricultural status quo. Their entrepreneurism not only directly helped the dissemination of NERICA but also pushed the Benin government toward policies for agricultural business development. In February 2008, the government issued a new agricultural development strategy plan aiming to establish an institutional, legal, regulatory, and administrative environment conducive to agricultural activities.

What can be learned from the NERICA case is that the dissemination of this new technology did not follow the conventional process of assistance programs and government adoption. There was a process of self-organization through various nongovernmental organizations. Self-motivated economic entrepreneurs started the process and propelled innovation. As a result, the private sector was able to push the government to adopt new policies that would be conducive to these innovations. These conditions then created more economic opportunities that drew more self-organized entrepreneurs to the program and thereby completed a healthy cycle of economic and technological improvement. This process as a whole can be understood as a "self-organizing system of innovation."

Industrial Clusters in Slovenia

Slovenia has made substantial economic progress since gaining its independence in 1991. At the end of the 1990s, Slovenia had a stable macroeconomic environment, with an average annual GDP growth rate of 4.3% and US$15,000 GDP per capita. Today Slovenia has nearly doubled that GDP per capita to US$27,000 and has jumped ahead of some of the "old" EU member state members such as Portugal and Greece. Despite favorable macroeconomic indicators, in the 1990s the economy depended on traditional industries with small profit margins and slow growth. Productivity was more than three times lower than the average productivity in the EU countries, while economic growth disproportionately depended on investments in physical assets, not knowledge or technology. Low education, weak public institutions, and insufficient social capital led to a shortage of competency, trust, and willingness to take risks.

In order to speed up the process of change and to stimulate business innovation, in 1999 the Slovenian Ministry of Economy launched an entrepreneurship and competitiveness policy.[17] Clustering was encouraged between similar and symbiotic firms as a way to increase knowledge creation and dissemination in key sectors. An initial mapping of potential clusters was conducted to analyze the geographic concentration of industries and the existing degree of networking and innovation systems, including linkages with universities, research centers, and other traditional centers of innovation.

Although the study revealed generally weak linkages and low levels of geographic concentration, 10 industries were nonetheless identified as having potential for cluster development. The cluster development program was articulated based on three interlinked measures: encouraging cooperation and networking between companies and R&D institutions; strengthening the knowledge, skills, and expertise required by key development actors (people and institutions) to promote the development and functioning of clusters; and forming clusters in practice.

The ministry began by co-financing projects involving companies and support institutions such as universities in the fields of marketing, product development, technology improvements, and specialization in supply chains. Three pilot clusters were supported with the objective of gaining knowledge and experience before any large-scale program was launched. The pilots followed a three-phase cluster development process: the initiation phase in which actors develop a common vision and devise an action plan for its implementation; an early growth phase when they implement the action plan and develop the platforms needed for the final phase; a final phase focused on R&D and internationalization. The model developed by pilots proved to be acceptable to the Slovenian environment, and a full-scale program was launched. Government financing was provided for the first year and then extended for two more years to those clusters with the best strategies. The government financial support was mainly used for R&D activities and training.

The government eventually supported 17 clusters. More than 400 firms and 100 business support institutions, universities, and research institutions participated with more than 66,500 employees. A total of 240 innovative projects have been launched as part of cluster initiatives in the areas of R&D between small and medium-sized enterprises (SMEs) and academic institutions, specialization in the value chain, internationalization, standardization, and training. At the beginning, the most important projects were focused on strengthening the cooperation between companies along the value chain and later on research and innovation. Almost all the clusters have internationalized and connected with foreign networks and clusters in less than two years. In 2006, the cluster program was completed. More demanding technology and research-oriented programs were launched to target specific technology, research, and science fields. Almost all of these post-2006 projects emerged from clustering activities.

Clusters provide crucial formal and informal linkages that increase trust among diverse actors, leading to greater exchange of individuals and ideas and key cooperation in areas no single firm or institution could achieve on its own. Despite advances in telecommunications, innovation in many sectors continues to be generated by and most easily transmitted between geographically proximate actors.

As farmer productivity is often constrained by lack of appropriate technology or access to best practice knowledge, inputs, and services, clusters may be able to provide pronounced benefits in the agro-sector. Certain types of clusters may have a more direct impact on poverty. These are the clusters in rural areas and in the urban informal economy; clusters that have a preponderance of SMEs, micro-enterprises, and home workers; clusters in labor-intensive sectors in which barriers to entry for new firms and new workers are low; and clusters that employ women, migrants, and unskilled labor.

In many African countries the agricultural sector is dominated by family-based small-scale planting. This structure slows down the diffusion and adoption of information and modern technology, a key driver of agricultural productivity and net growth. One of the main challenges is therefore to enhance technology transfer from knowledge producers to users in the rural regions where small-scale household farming dominates. Clusters can overcome these shortcomings by creating the linkages and social capital needed to foster innovation and technology transfer. However, clusters are not a cure-all for African agricultural innovation, and we must therefore look closely at the conditions under which clusters can work, the common stages of their development, and key factors of their success.

Clusters cannot be imposed on any landscape. They are most likely to form independently or to succeed once seeded by government when they are collocated with key inputs, services, assets, and actors. Clusters are most likely to form and succeed in regions that already possess the proper input, as well as in industries that have a dividable production process and a final product that can be easily transported. Clusters are also more likely in knowledge or technology-intensive businesses (like agriculture), where breakthroughs can instigate quick and significant increases in productivity. Clusters also benefit from preexisting tightly knit social networks, which provide fertile ground for more complex knowledge generation and sharing infrastructure.

Policies for Cluster Development

Cluster development could benefit from the experiences outlined above. In the first phase, governments should lead the formation of clusters by identifying strategic regions with the right human, natural, and institutional resources to establish a competitive advantage in a key sector. Governments can then nurture a quick flow of investment, ideas, and even personnel from the public sector to private firms. As government-funded initiatives deliver proof of concept, governments should make way for private enterprise and give up their ownership stakes in the burgeoning agro-industries they helped create.

As government involvement decreases, clusters move to formalize the connections between key actors through producer associations and other cooperative organizations. Strong bonds formed in the early phases of cluster formation allow diverse actors to come together on common sets of standards in key areas of health, safety, and environment. Quality control and enhanced production are critical for clusters to move beyond their local markets and into more lucrative national or international export markets. Despite their decreasing role, governments can continue to play a key part in this process by putting in place regulations that ease, rather than obstruct, firms' efforts to meet complex international health, environmental, and labor standards.

This strong foundation in place, clusters can move to additional cooperative efforts focused on international marketing and export, and complex partnerships with large multinational companies. Firms can band together to accomplish what none of them can do individually: achieve national and international brand recognition.

Innovation systems likewise cannot be imposed by outside actors and must have substantial buy-in from local government, business groups, and citizen groups. Additionally, governments must wrestle with the possibility that although clusters enhance knowledge generation and transmission within themselves, strong social and practical connections within clusters may actually make communications between them less likely.[18] Linkages between clusters are therefore critical, and this is an area in which regional organizations can play a particularly important role.

Local governments played a critical role in determining initial potential for clustering by evaluating natural and human resources, already existing clusters, and markets in which their area might be able to deliver a competitive advantage. Local governments also assessed and in many cases fueled popular citizen, business, and public institutional support for enhanced cooperation, a key precursor for clusters. As clusters depend on physical and cultural proximity to encourage knowledge creation and sharing, local governments can encourage these exchanges between firms, individual producers, NGOs, and research and academic institutions even before funding has been set aside for a specific cluster.

While local authorities are best placed to determine the potential for clusters in specific areas, national governments may be better positioned (particularly in Africa) to provide the financial and regulatory support necessary for successful clusters. National governments use state-owned banks, tax laws, and banking regulations to encourage loans to businesses and organizations in these key clusters. They also help finance investments by constructing key infrastructure, including ports, roads, and telecommunications. Finally, governments play a key role in responding to pressure from the clusters to create regulatory frameworks that help them to meet stringent international environmental, health, and labor standards. National governments can also play a central role in convincing nationally funded research and academic institutions to participate actively in clusters with businesses and individual producers.

While clusters lower barriers to knowledge creation and sharing within themselves, the opposite may be true across different national or regional economic activities. This isolation may limit innovation within clusters, or worse could lead to negative feedback cycles based on the phenomenon of "lock-in," whereby clusters increasingly focus on outdated or noncompetitive sectors or strategies.[19] Regional institutions and linkages can play a key role in making and maintaining these external links by supporting the exchange of information, and in particular personnel, between clusters. In Africa, regional institutions could also support the idea of regional centers of excellence based around key specialties—for example, livestock in East Africa.

The Role of Local Knowledge

Strengthening local innovation systems or clusters will need to take into account local knowledge, especially given emerging concerns over climate change.[20] Farming communities have existed for a millennium, and long before there were modern agricultural innovations, these communities had to have ways to manage their limited resources and keep the community functioning. Communities developed local leadership structures to encourage participation and the ideal use of what limited resources were available. In the past few centuries, colonial intervention and the push for modern methods have often caused these structures to fail as a result of neglect or active destruction. However, these traditional organizational mechanisms can be an important way to reach a community and cause its members to use innovations or sustainable farming techniques.[21]

While governments and international organizations often overlook the importance of traditional community structures, they can be a powerful tool to encourage community members in the use of new technologies or the revival of traditional methods that are now recognized as more effective.[22] Communities retain the knowledge of and respect for these traditional leadership roles and positions in a way that outside actors can not, and they will often adopt them as a way to manage community agricultural practices and learning. It is this place-based innovation in governance that accounts to a large extent for institutional diversity.[23]

India's recent reintroduction of the Vayalagams as a means of water management serves as a good example of how traditional systems can still serve the local communities in which they originated as a means of agricultural development and economic sustainability. A long-standing tradition in India in the pre-colonial period was the use of village governance structures called Vayalagams to organize and maintain the use of village water tanks. These tanks were an important component of rain-fed agriculture systems and provided a reservoir that helped mitigate the effects of flooding and sustain agriculture and drinking-water needs throughout the dry season by capturing rainwater.

The Vayalagams were groups of community leaders who managed the distribution of water resources to maximize resources and sustainability, and to ensure that the whole community participated in, and benefited from, the appropriate maintenance of the tanks. Under British colonial rule, and later under the independent Indian government, irrigation systems became centralized and communities were no longer encouraged to use the tanks, so both the physical structures and the organizations that managed them fell into disrepair.

As the tank-fed systems fell apart and agricultural systems changed, rural communities began to suffer from the lack of sufficient water to grow crops. One solution to this problem has been to revitalize the Vayalagam system and to encourage the traditional community networks to rebuild the system of tanks. Adopting traditional methods of community organization has tapped into familiar resources and allowed the Development of Humane Action (DAHN) Foundation—an Indian NGO—to rally community ownership of the project and gain support for rebuilding the system of community-owned and managed water tanks. The tanks were a defunct system when the DHAN Foundation incorporated in 2002. Now, the Tank-Fed Vayalagam Agricultural Development Programme works in 34 communities and has implemented 1,807 micro finance groups that comprise 102,266 members. It funds the program with a 50% community contribution and the rest from the foundation. This redeployment of old community organizations has resulted in rapid proliferation of ideas and recruitment of farmers.

Reforming Innovation Systems

As African countries seek to promote innovation regionally, they will be forced to introduce far-reaching reforms in their innovation systems to achieve two important goals. The first will be to rationalize their research activities in line with the goals of the Regional Economic Communities (RECs). The second will be to ensure that research results have an impact on the agricultural productive sector. Many emerging economies have gone through such reform processes. China's reform of its innovation system might offer some insights into the challenges that lie ahead.

Partnerships between research institutes (universities or otherwise) and industries are crucial to encourage increased research and promote innovation. Recent efforts in China to reform national innovation systems serve to demonstrate the importance of "motivating universities and research institutes (URIs), building up the innovative capacities of enterprises, and promoting URI-industry linkages."[24] Before the most recent reforms, China's model mimicked the former Soviet Union's approach to defense and heavy industry R&D, in which the system was highly centralized. Reforms allowed for increased flexibility, providing incentives to research institutes, universities, and business enterprises to engage in research. The case study of China's science and technology (S&T) reforms demonstrates the efficacy of using policy and program reform to increase research, patents, publications, and other innovations.

During the pre-reform period from 1949 to the 1980s, China focused on military research, carried out for the most part by public research institutes and very sparingly by universities. Almost all research was planned and funded by the government with individual enterprises (which often had their own S&T institutes and organizations) engaging in little to no research and development.

With the hope of developing the country through education and research, China created the slogan "Building the nation through science and education" to underscore their 1985 reforms. Efforts were made to increase university and research institution collaboration with related business industries, and in the 1990s this was furthered by motivating universities and institutions to establish their own enterprises.

Reforms occurred in three stages, the first of which spanned 1985 to 1992. Here, the government initiated reform by encouraging universities and research institutes (URIs) to bolster their connections with industry—one method used was to steeply cut the research budget for universities and other institutes with the goal of causing the URIs to turn to industry for support and thus facilitate linkages and partnerships. Additional laws and regulations regarding patent and technology transfer were passed, and by the end of 1992, 52 high-tech development zones had been set up, with 9,687 enterprises and a total turnover of renminbi (RMB) 56.3 billion.

From 1992 to 1999, the second stage of reform saw the creation of the "S&T Progress Law" and the "Climbing Program" to encourage research as well as the increased autonomy regarding research given to URIs. A breakthrough that strengthened partnerships between URIs and industry was the 1991 endorsement of enterprises that were affiliated with URIs. Linkages that were encouraged included technical services, partnerships in development, production, and management, as well as investment in technology. Vast improvements were seen immediately: from 1997 to 2000, university-affiliated enterprises experienced average annual sales income growth of 32.3%, with 2,097 high-tech ones emerging in China with a total net worth of US$3.8 billion by 2000.

During the third stage, starting in 1999, China sought to both strengthen the national innovation systems and facilitate the commercialization of R&D results. One key measure was the transformation of state-owned applied research institutes into high-tech firms or technical service firms. Of 242 research institutes that were to be transformed from the former State Committee for Economics and Trade, 131 merged with corporations (groups), 40 were transformed into S&T corporations under local governments, 29 were transformed into large S&T

corporations owned by the central government, 18 were transformed into agencies, and the remaining 24 turned into universities or were liquidated. A total of 1,149 transformations were carried out by the end of 2003.

New policies and programs helped bring about changes during the reform period. The "Resolution on the Reform of the S&T System," released in 1985, aimed to improve overall R&D system management, including encouraging research personnel mobility and integration of science and technology into the economy through the introduction of flexible operating systems. Peer review of projects and performance brought about a degree of transparency. Reform policies promoted more flexible management of R&D, technology transfer, linkages between URIs and industry, and commercialization of high-tech zones.

The many new programs were meant to serve different purposes and have been shown to be effective in general. One particular program was extremely important in the high-tech area. The "863 Program," which was launched in 1986, sought to move the country's overall R&D capacity to cutting-edge frontiers in priority areas such as biotechnology, information, automation, energy, advanced materials, marine, space, laser, and ocean technology. Another goal of the 863 Program was to promote the education and training of professionals for the 21st century by mobilizing more than 10,000 researchers for 2,860 projects every year. An example of another program was "The Torch Program," launched in 1988. By reducing regulation, building support facilities, and encouraging the establishment of indigenous high-tech firms in special zones, the program aimed to establish high-tech firms. Success is evident: "From 1991 to 2003, 53 national high-tech zones had been established" especially in the information technology, biotechnology, new materials, and new energy technologies industries. "The national high-tech zones received RMB 155 billion investments in infrastructure and hosted 32,857 companies in 2003."[25] It appears that these early but critical reform efforts have put China on a path that could enable it to catch up with the industrialized countries in science and innovation.[26]

Because the ultimate goal of the science and technology reforms in China were meant to strengthen national innovation systems and promote innovation activities among the key players in the system, it was necessary for URIs, industry, and the government to interact. The impact of the reforms is seen in the stark contrast between the years 1987 and 2003. In 1987, government-funded public research institutes dominated R&D research, with universities carrying out education and enterprises involved in restricted innovations in "production and prototyping." For this reason, URIs found no reason to conduct applied research or to commercialize their research results.

By 2003, R&D expenditure had risen by more than eightfold. Most distinctive was the large increase in R&D units, employees, and expenditures of enterprises. This was brought about in part from the transformation of 1,003 or 1,149 public research institutes into enterprises or parts of enterprises. Additionally, after the 1991 endorsement of university-affiliated enterprises, a great expansion occurred such that by 2004, 4,593 of them existed with annual income of RMB 97 billion. Another factor was increased competition that created incentive to engage in R&D. Finally, in general the R&D potential of the firms has increased as a result of the more supportive environment resulting from S&T reform.

The increased R&D expenditure from enterprises demonstrates the overall success of the science and technology reforms. This success is also seen in the improved URI-industry linkages, as is shown in the decrease in government spending from 79% in 1985 to 29.9% in 2000. URIs (either transformed or public ones) have forged close links with the private sector "through informal consulting by university researchers to industry, technology service contracts, joint research projects, science parks, patent licensing, and URI-affiliated enterprise."[27] Another success from the S&T reform is the great increase in patents from domestic entities as well as the larger number of publications.

Science and technology reform in China demonstrated the importance of creating linkages between industry and institutes for research and education. Despite the great success, there are a few cautionary lessons to be learned from China's actions. For example, while great improvements were found in the linkages between URIs and industries, there has been a lack of focus on science and technology administration. Because the many governmental and nongovernmental bodies work independently, there is a danger of inefficiency in the form of redundancy or misallocation of R&D resources. The reform's focus on commercializing S&T has also prevented further development of basic research and other research aimed at public benefit (with such research stuck at 6% of all research funding). A final concern is the controversy surrounding university-affiliated enterprises that emphasize the operation, ownership structure, and the de-linking of such enterprises from their original parent universities. Critics believe that commercial goals may hinder other university mandates about pure academics. When creating comprehensive reform of such magnitude, one must be careful to take into account these potential issues.

China's science and technology reforms demonstrate the potential for expanding research by supporting the formation of URI-industry partnerships and linkages. The benefits are clear and developing countries should greatly consider using China's case as a model for the establishment of similar programs and policies.

African countries can rationalize their research activities through an entity that can draw lessons from the Brazilian Agricultural Research Corporation (EMBRAPA). This successful institutional innovation was designed to respond to a diversity of agricultural needs over a vast geographical area. It has a number of distinctive features that include the use of a public corporation model; national scale of operations in nearly all states; geographical decentralization; specialized research facilities (with 38 research centers, 3 service centers, and 13 central divisions); emphasis on human resource development (74% of 2,200 researchers have doctoral degrees while 25% have master's training); improvements in remuneration for researchers; and strategic outlook that emphasizes science and innovation as well as commercialization of research results.[28]

Conclusion

Agricultural innovation has the potential to transform African agriculture, but only if strong structures are put in place to help create and disseminate critical best practices and technological breakthroughs. In much of Africa, linkages between farmers, fishermen, and firms and universities, schools, and training centers could be much stronger. New telecommunications technologies such as mobile phones have the potential to strengthen linkages, but cluster theory suggests that geography will continue to matter regardless of new forms of communication. Groups that are closer physically, culturally, and socially are more likely to trust one another, exchange information and assets, and enter into complex cooperative production, processing, financing, marketing, and export arrangements.

Local, national, and regional authorities must carefully assess where clusters may prove most successful and lay out clear plans for cluster development, which can take years if not decades. Local authorities should focus on identifying potential areas and industries for successful clusters. National governments should focus on providing the knowledge, personnel, capital, and regulatory support necessary for cluster formation and growth. And regional authorities should focus on linking national clusters to one another and to key related global institutions. Throughout these processes, public and private institutions must work cooperatively, with the latter being willing to transfer knowledge, funding, and even personnel to the private sector in the early stages of cluster development.

To promote innovation, the public sector could further support interactions, collective action, and broader public-private partnership programs. The country studies suggest that from a public sector perspective, improvements in agricultural innovation system policy design, governance, implementation, and the enabling environment will be most effective when combined with activities to strengthen innovation capacity. Success stories in which synergies were created by combining market-based and knowledge-based interactions and strong links within and beyond the value chain point to an innovation strategy that has to be holistic in nature and focus, in particular, on strengthening the interactions between key public, private, and civil society actors.

4

ENABLING INFRASTRUCTURE

Enabling infrastructure (public utilities, public works, transportation, and research facilities) is essential for agricultural development. Infrastructure is defined here as facilities, structures, associated equipment, services, and institutional arrangements that facilitate the flow of agricultural goods, services, and ideas. Infrastructure represents a foundational base for applying technical knowledge in sustainable development and relies heavily on civil engineering. This chapter outlines the importance of providing an enabling infrastructure for agricultural development.[1] Modern infrastructure facilities will need to reflect the growing concern over climate change. In this respect, the chapter will focus on ways to design "smart infrastructure" that takes advantage of advances in the engineering sciences as well as ecologically sound systems design. Unlike other regions of the world, Africa's poor infrastructure represents a unique opportunity to adopt new approaches in the design and implementation of infrastructure facilities.

Infrastructure and Development

Poor infrastructure and inadequate infrastructure services are among the major factors that hinder Africa's sustainable development. This view has led to new infrastructure development approaches.[2] Without adequate infrastructure, African countries will not be able to harness the power of science and innovation to meet sustainable development objectives and be competitive in international markets. Roads, for example, are critical for supporting rural development. Emerging evidence suggests that in some cases low-quality roads have a more significant impact on economic development than high-quality roads. In addition, all significant scientific and technical efforts require reliable electric power and efficient logistical networks. In the manufacturing and retail sectors, efficient transportation and logistical networks allow firms to adopt process and organizational innovations, such as the just-in-time approach to supply chain management.

Infrastructure promotes agricultural trade and helps integrate economies into world markets. It is also fundamental to human development, including the delivery of health and education services. Infrastructure investments further represent untapped potential for the creation of productive employment. For example, it has been suggested that increasing the stock of infrastructure by 1% in an emerging country context could add 1% to the level of GDP. But in some cases the impact has been far greater: the Mozal aluminum smelter investment in Mozambique not only doubled the country's exports and added 7% to its GDP, but it also created new jobs and skills in local firms.

Reducing public investment in infrastructure has been shown to affect agricultural productivity. In the Philippines, for example, reduction in investment in rural infrastructure led to reductions in agricultural productivity.[3] This decline in investment was caused by cutbacks in agricultural investments writ large, as well as by a shift in focus from rural infrastructure and agricultural research to agrarian reform, environment, and natural resource management. Growth in Philippine agriculture in the 1970s was linked to increased investments in infrastructure, just as declines in the same sector in the 1980s were linked to reduced infrastructure investment (caused by a sustained debt crisis).

Evidence from Uganda suggests that public investment in infrastructure-related projects has contributed significantly to rural development.[4] Uganda's main exports are coffee and cotton; hence, the country depends heavily on its agricultural economy. Political and economic turmoil in the 1970s and 1980s in Uganda led to the collapse of the economy and agricultural output. Reforms in the late 1980s allowed Uganda to improve its economic growth and income distribution. In spite of economic growth ranging between 5% and 7%, the growth of the agricultural sector has been very low, averaging 1.35% per annum. Even if the Ugandan government has made great strides in welfare improvement, the rural areas still remain relatively poor. In addition, due to the disparity between male and female wages in agriculture, women are more affected by poverty than men.

The Ugandan government has been spending on a wide variety of sectors including agriculture, research and development, roads, education, and health (data in other sectors such as irrigation, telecommunications, and electricity are limited). Previous studies have mostly measured the effectiveness of government spending based on budget implementation.

Government spending on agricultural research and extension improved agricultural production substantially in Uganda. Growth in agricultural labor productivity, rural wages, and nonfarm employment have emerged as important factors in determining rural poverty, so much so that the public expenditure on agriculture outweighs the education and health effect. Investment in agriculture has been shown to increase food production and reduce poverty. Roads linking rural areas to markets also serve to improve agricultural productivity and increase nonfarm employment opportunities and rural wages.

Having a high HIV/AIDS prevalence, a large share of Uganda's health expenditure goes toward prevention and treatment. Despite the high expenditure in health services, there does not seem to be a high correlation between health expenditure and welfare improvement.

Infrastructure and Agricultural Development

Transportation

Reliable transportation is absolutely critical for growth and innovation in African agriculture and agribusiness. Sufficient roads, rail, seaports, and airports are essential for regional trade, international exports, and the cross-border investments that make both possible. Innovation in other areas of agriculture such as improved genetic material, better access to capital, and best farming practices will produce results only if farmers and companies have a way to get their products to market and get critical inputs to farms.

Transportation is a key link for food security and agribusiness-based economic growth. Roads are the most obvious and critical element, but modern seaports, airports, and rail networks are also important, particularly for export-led agricultural innovation such as cut flowers or green beans in Kenya, neither of which would be possible without an international airport in Nairobi. To that end, many African countries have reprioritized infrastructure as a key element in their agricultural development strategies. This section will examine the role roads have played in China's rural development and poverty alleviation, as well as two cases in African transportation investment: Ghana's rural roads project and Mali's Bamako-Sénou airport improvement project.

Ghana's Rural Roads Project is expected to open new economic opportunities for rural households by lowering transportation costs (including travel times) for both individuals and cargo to markets and social service delivery points. The project will include new construction, as well as the improvement of over 950 kilometers of feeder roads, which, along with the trunk roads, will benefit a total population of more than 120,000 farming households with over 600,000 members. These activities will increase annual farm incomes from cultivation by US$450 to about US$1,000. For many of the poor, the program will represent an increase of one dollar or more in average income per person per day. In addition to sparking growth in agriculture, the feeder roads will also help facilitate transportation linkages from rural areas to social service networks (including, for instance, hospitals, clinics, and schools).

The Airport Improvement Project will expand Mali's access to markets and trade through improvements in the transportation infrastructure at the airport, as well as better management of the national air transport system. However, Mali is landlocked and heavily dependent on inadequate rail and road networks and port facilities in countries whose recent instability has cost Mali dearly. Before the outbreak of the Ivorian crisis, 70% of Malian exports were leaving via the port of Abidjan. In 2003, this amount dwindled to less than 18%. Mali cannot control overland routes to international and regional markets. Therefore, air traffic has become Mali's lifeline for transportation of both passengers and export products.

Malian exports are predominantly agriculture based and depend on rural small-scale producers, who will benefit from increased exports in high-value products such as mangoes, green beans, and gum arabic. The Airport Improvement Project is intended to remove constraints to air traffic growth and increase the airport's efficiency in both passenger and freight handling through airside and landside infrastructure improvements, as well as the establishment of appropriate institutional mechanisms to ensure effective management, security, operation, and maintenance of the airport facilities over the long term.

In response to requirements for safety and security audits by the International Civil Aviation Organization and the United States Federation Aviation Administration, Mali is in the process of restructuring and consolidating its civil aviation institutional framework. One major result has been the establishment of the new civil aviation regulatory and oversight agency in December 2005, which now has financial and administrative independence. The Airport Improvement Project will reinforce the agency by providing technical assistance to establish a new organizational structure, administrative and financial procedures, staffing and training, and provision of equipment and facilities. Additionally, the project will rationalize and reinforce the airport's management and operations agency by providing technical assistance to establish a model for the management of the airport and the long-term future status and organization of agency.

Since 1985 the government of China has given high priority to road development, particularly the construction of high-quality roads such as highways and freeways. While the construction of high-quality roads has taken place at a remarkably rapid pace, the construction of lower-quality and mostly rural roads has been slower. Benefit-cost ratios for lower-quality roads (mostly rural) are about four times *larger* than those for high-quality roads when the benefits are measured in terms of national GDP.[5]

In terms of welfare improvement, for every yuan invested, lower-quality roads raised far more rural and urban poor people above the poverty line than did high-quality roads. Without these essential public goods, efficient markets, adequate health care, a diversified rural economy, and sustainable economic growth will remain elusive. Effective development strategies require good infrastructure as their backbone. The enormous benefit of rural roads in China likely holds true for other countries as well. Investment in rural roads should be a top priority for reducing poverty,

maximizing the positive effects of other pro-poor investments, and fostering broadly distributed economic growth.
Although highways remain critical, lower-cost, often lower-
quality rural feeder roads are of equal and in some cases even
greater importance.

As far as agricultural GDP is concerned, in today's China
additional investment in high-quality roads no longer has a
statistically significant impact while low-quality roads are not
only significant but also generate 1.57 yuan of agricultural
GDP for every yuan invested. Investment in low-quality roads
also generates high returns in rural nonfarm GDP. Every yuan
invested in low-quality roads yields more than 5 yuan of rural
nonfarm GDP. Low-quality roads also raise more poor people
out of poverty per yuan invested than high-quality roads,
making them a win–win strategy for growth in agriculture
and poverty alleviation. In Africa, governments can learn
from the Chinese experience and make sure their road
programs give adequate priority to lower-quality and rural
feeder roads.

Irrigation

Investment in water management is a crucial element of successful agricultural development and can be broken into two
principal areas: policy and institutional reforms on the one
hand and investment, technology, and management practices
on the other. Water is also a critical input beyond agriculture,
and successful irrigation policies and programs must take into
account the key role of water in energy production, public health,
and transportation. For small farmers, low-cost technology
is available, and there are cost-efficient technical solutions
in even some of Africa's most difficult and arid regions. Despite
the availability of these technologies, Africa has not seen widespread adoption of these techniques and technologies. Part of
the problem is the availability of finance and the slow spread

of knowledge, but equally important is the role of government regulations and subsidies.

Successful strategies for improved water management and irrigation must therefore not only focus on new technologies but also on creating policies and regulations that encourage investment in irrigation, not just at the farm but also at the regional level. Access to reliable water supplies has proven a key determinant not just in the enhancement of food security but also in farmers' ability to climb higher up the value chain toward cash crops and processed foods. Innovative farmers involved in profitable agro-export may represent a new constituency for the stewardship of water resources, as they earn significantly higher incomes per unit of water than conventional irrigators. Our analysis will focus first on innovation in water management practices, technology, and infrastructure (including examples from Mali, Egypt, and India). In the final section of this chapter, we will also address key water policy and institutional reforms necessary to create an environment in which governments, international institutions, NGOs, and private businesses will be encouraged to make investments in irrigation infrastructure.

Begun in 2007, the Alatona Irrigation Project will provide a catalyst for the transformation and commercialization of family farms, supporting Mali's national development strategy objectives to increase the contribution of the rural sector to economic growth and help achieve national food security. Specifically, it will increase production and productivity, improve land tenure security, modernize irrigated production systems and mitigate the uncertainty from subsistence rain-fed agriculture, thereby increasing farmers' incomes. The Alatona Irrigation Project will introduce innovative agricultural, land tenure, credit, and water management practices, as well as policy and organizational reforms aimed at realizing the Office du Niger's potential to serve as an engine of rural growth for Mali. This project seeks to develop 16,000 hectares of newly irrigated lands in the

Alatona production zone of the Office du Niger, representing an almost 20% increase of "drought-proof" cropland.

Egypt depends almost entirely on 55.5 billion cubic meters per year of water from the Nile River. This allocation represents 95% of the available resource for the country. Approximately 85% of the Nile water is used for irrigation. Demand for water is growing while the options for increasing supply are limited. To respond, the Ministry of Water Resources and Irrigation (MWRI) has been implementing an Integrated Water Resource Management (IWRM) Action Plan. Its key strategy is to improve demand management. The Integration Irrigation Improvement and Management Project (IIIMP), a part of MWRI, has been implementing the IWRM Action Plan.

The IIIMP adopted a three-point strategy: (1) proper sizing of the improved infrastructure to optimize capital costs; (2) technical innovations to increase cost savings and functionality; and (3) extension of the improvement package to the whole system (including tertiary and on-farm improvements). The IIIMP Project appraisal document envisaged (1) an average increase in farmers' annual income of approximately 15%, (2) water savings of approximately 22%, and (3) an overall economic rate of 20.5%. In the pilot areas where the program was implemented, yields increased by 12% to 25%. Net incomes per cultivated acre are increasing by 20% to 64% as a result of the combined effects of increased productivity and reduction of the irrigation costs (depreciation and operations and maintenance of pumps.

Sugarcane cultivation requires significant water resources, but in much of India it has been cultivated using surface irrigation, where water use efficiency is very low (35%–40%), owing to substantial evaporation and distribution losses.[6] A recent study of sugarcane cultivation in Tamil Nadu, India, has shown that using drip irrigation techniques can increase productivity by approximately 54% (30 tons per acre) and cut water use by approximately 58% over flood irrigation. Unlike surface methods of irrigation, under drip methods, water is supplied directly to the root zone of the crops through a network of pipes, a system that saves enormous amounts of water by reducing evaporation and distribution losses. Since water is supplied only at the root of the crops, weed problems are less severe and thus the cost required for weeding operations is reduced significantly. The system also requires little if any electricity.

Although new and larger studies are necessary, initial analysis suggests that investment in drip irrigation in Indian sugarcane cultivation is economically viable even without subsidy and may also be applicable in Africa where many farmers have no or limited access to electricity-powered irrigation, water resources are increasingly threatened by climate change and environmental degradation, and less than 4% of the arable land is currently irrigated.

Further, the present net worth indicates that in many cases farmers can recover their entire capital cost of drip irrigation from first-year income without subsidy. Despite these gains, two impediments must be overcome for drip irrigation to be more widely used not just in India, but in much of the developing world. First, too few farmers are aware of the availability and benefits of drip irrigation systems, which should be demonstrated clearly and effectively through a quality extension network. Second, despite the quick returns realized by many farmers using drip irrigation, the systems require significant capital up front. Banks, microcredit institutions, companies, and governments will need to consider providing credit or subsidies for the purchase of drip irrigation.

The total cultivated land area of the Common Market for Eastern and Southern Africa (COMESA) amounts to some 71.36 million hectares. Of this only about 6.48 million hectares are irrigated, representing some 9% of the total cultivated land area. Besides available land area for irrigation, the region possesses enormous water resources and reservoir development potential to allow for expansion. Of the world's total of 467 million hectares of annualized irrigated land areas, Asia accounts for 79 percent (370 million hectares), followed by Europe (7%) and North America (7%). Three continents—South America (4%), Africa (2%), and Australia (1%)—have a very low proportion of global irrigation. COMESA could contribute significantly to agricultural food production and poverty alleviation through expanding the land under irrigated agriculture and water management under rain-fed farming to effect all year round crop and livestock production.

COMESA has recently made assessments through the Comprehensive Africa Agriculture Development Programme (CAADP) stocktaking reports involving some representative countries with respect to agriculture production options and concluded that regional economic growth and food security could be accelerated through investment in irrigation and agriculture water management. Agriculture water-managed rain-fed yields are similar to irrigated yields and always higher than rain-fed agriculture yields. This scenario builds a watertight case for promoting or expanding irrigated land in COMESA.

The best solution to poverty and hunger alleviation is to provide people with the means to earn income from the available resources they have. Small-scale irrigation development coupled with access to long-term financing, access to markets, and commercial farming expertise by producers will go a long way in achieving food security and overall economic development. COMESA has created an agency called the Alliance for Commodity Trade in Eastern and Southern Africa (ACTESA) to implement practical investment actions by engaging public private sector partnerships. In the areas of irrigation and agriculture water management, COMESA has begun to implement a number of important activities, described next.

Accelerated adoption of appropriate small-scale irrigation technologies and improved use and management of agriculture water will facilitate increased agricultural production and family incomes. The rain-fed land area will require agriculture water management strategies such as conservation agriculture, which enhances production. Appropriate investment in field systems for irrigation with modest investments will help smallholder farmers adopt irrigation technology whereas the majority who practice rain-fed agriculture would improve agriculture productivity by managing rainwater through systems such as conservation agriculture technology. COMESA is embarking on reviewing the policy and legal framework in water resources management programs including transboundary shared water resources management policies under CAADP. This will include actions toward adaptation by member states of regional water resources management policies.

COMESA is working with regional and international organizations such as Improved Management of Agriculture Water in Eastern and Southern Africa (IMAWESA), East African Community (EAC), Southern African Development Coordination (SADC), Interovernmental Authority for Development (IGAD), International Water Management Institute (IWMI), and Wetland Action-UK in creating awareness in regional sustainable water resources management by creating and strengthening water dialogue platform and communication strategies. Through ACTESA, COMESA will help develop regional water management information systems observation networks so as to enhance mapping for water harvesting resources and water utilization in COMESA.

To realize the benefits of irrigation and agriculture water management, COMESA is promoting investment in the following areas: reservoir construction for storage of water to command an expansion of land area under irrigation by 30% in five years; inland water resources management of watershed basins in the COMESA region including, policy and legal frameworks in trans-boundary shared water resources management, harmonizing shared water resources policies to optimize utilization, strengthening regional institutions involved in water resources management, and establishment of a regional water resources management information system; building capacity and awareness for sustainable water resources utilization and management for agricultural food production; rapid expansion of terraces for hilly irrigation in some member states; and promotion and dissemination of appropriate irrigation and agriculture water management technology transfer and adoption. These include smallholder irrigation infrastructure.

Energy

To enhance agricultural development and to make progress in value-added agro-processing, Africa needs better and more consistent sources of energy. Rolling blackouts are routine in much of west, central, and eastern Africa, and much of Africa's power generation and transmission infrastructure needs repair or replacement. What Africa lacks in adequate deployment, however, it makes up for in potential. Africa is endowed with hydro, oil, natural gas, solar, geothermal, coal and other resources vast enough to meet all its energy needs. Nuclear energy is also an option. The hydro potential of the Democratic Republic of the Congo is itself enough to provide three times as much power as Africa currently consumes.

The first step to improved power generation and transmission is to repair and upgrade Africa's existing energy infrastructure. Many African countries are operating at less than half their installed potential due to inadequate maintenance and operation. Connecting rural areas to national grids can in some cases be cost prohibitive, so governments must also look for innovative solutions such as wind, solar, biomass, and geothermal to provide power at the small farm level. Finally, while countries will undoubtedly look first within their own borders for resources, advanced energy planning should also consider that the most affordable and reliable power may be in neighboring states. Large power generation schemes may also require cooperative agreement on resource management and funding from a host of African and international sources. Cross-border energy networks could help create a common market for energy, spur investment and competition, and lead to a more efficient path of enhanced energy infrastructure.

An example of such a regional energy system is the West African Power Pool (WAPP). Under an agreement signed by 14 ECOWAS members in 2000, countries plan to develop energy production facilities and interconnect their respective electricity grids. According to the agreement, the work would be approached in two phases. ECOWAS estimates that 5,600 kilometers of electricity lines connecting segments of national grids will be put in place. About US$11.8 billion will be needed for the necessary power lines and new generating plants. This infrastructure would give the ECOWAS subregion an installed capacity of 10,000 megawatts and, critically for agro-processing and business investment, dramatically increase not just the amount but also the reliability of electricity in west Africa.

The key objectives of the WAPP are to establish a well-functioning, cooperative, power-pooling mechanism among national power utilities of ECOWAS member states, based on a transparent and harmonized legal, policy, regulatory, and commercial framework. This framework would promote cross-border exchange of electricity on a risk-free basis; assure national power utilities of mutual assistance to avoid a regional power system collapse, or rapid restoration of interconnected regional power; reduce collective vulnerability of ECOWAS member states to drought-induced power supply disruptions; give ECOWAS member states increased access to more stable and reliable supplies of electricity from lower-cost regional sources of (hydro and gas-fired thermal) power generation; and create clear and transparent pricing arrangements for cross-border trade to facilitate electricity exchange and trade.

The WAPP organization has been created to integrate the national power system operations into a unified regional electricity market—with the expectation that such a mechanism would, over the medium to long term, assure the citizens of ECOWAS member states a stable and reliable electricity supply at affordable costs. This will create a level playing field facilitating the balanced development of the diverse energy resources of ECOWAS member states for their collective economic benefit, through long-term energy sector cooperation, unimpeded energy transit, and increased cross-border electricity trade. The major sources of electricity under the power pool would be hydroelectricity and gas to fuel thermal stations. Hydropower would be mainly generated on the Niger (Nigeria), Volta (Ghana), Bafing (Mali), and Bandama (Côte d'Ivoire) rivers. The World Bank has committed a $350-million line of credit for the development of the WAPP, but a billion more is needed in public and private financing.

Most of the power supply in Africa is provided by the public sector. There is growing interest in understanding the ability of independent power projects (IPPs) in Africa by evaluating a project's ability to produce reliable and affordable power as well as reasonable returns on investment.[7] In the context of their individual markets the 40 IPPs under consideration here have played a complementary role to state-owned power projects, filling gaps in supply. It was also hoped that, once established, these private entities would introduce competition into the market.

Evidence suggests that there is a dichotomy between relatively successful IPPs situated mainly in the northern African nations of Egypt, Morocco, and Tunisia, and the sub-Saharan examples including Ghana, Kenya, Nigeria, and Tanzania, which have been less successful. A wide variety of country-level factors including investment climate, policy frameworks, power sector planning, bidding processes, and fuel prices all impacted outcomes for these various IPPs. Despite their private nature, ultimately, it is the perceived balance of commitment between sponsors and host-country governments that plays one of the largest roles in the outcome of the IPP. A leading indicator of imbalance is frequent and substantial contract changes.

The presence of a favorable climate for investment influenced the outcome of the IPPs. In the more successful northAfrican examples, Tunisia carried an investment grade rating, while Egypt and Morocco were both only one grade below investment grade. In contrast, of those nations located in sub-Saharan Africa, none received an investment grade rating. The great demand for IPPs in Africa at the time meant that those with superior investment profiles were able to attract more investors and had a basis for negotiating a more balanced contract.

Few of the nations in question have established a clear and coherent policy framework within which an IPP could sustainably operate. The soundest policy frameworks again are found in the north, with Egypt, which contains 15 IPPs, being the strongest. This policy framework features a clearly defined government agency in the Egyptian Electricity Authority, which has authority over the procurement of IPPs, the allocation of new generation capacity, and the ability to set benchmarks to increase competition among public facilities. Kenya set itself apart in this context with the establishment of an independent regulator—the Electricity Regulatory Board—which has helped to significantly reduce power purchase agreement charges, set tariffs, and mediate the working relationship between the public and private sectors. Evidence suggests that if a regulator is established prior to negotiation of the IPP, and acts in a transparent, fair, and accountable manner, this office can have a significantly positive effect on the outcomes for the host country and investor.

A coherent power sector plan follows from a strong policy framework and includes setting a reliability standard for energy security, supply and demand forecasts, a least-cost plan, and agreements on how new generation will be divided between public and private sectors. It is equally important that these functions are vested in one empowered agency. Failure to meet these goals is apparent in the examples of Tanzania (Songo Songo), Kenya (Westmont plant, Iberafrica plant), Nigeria (AES Barge), and Ghana, which fast-tracked IPPs to meet intermediate power shortages in the midst of drought conditions. The results were unnecessary costs and time delays for all, and in the case of the Nigerian and Ghanaian facilities, an inability to efficiently establish power purchase agreements.

The main lesson learned here is that without a strong legislative foundation and coherent planning, contracts were unlikely to remain intact. Instability of contracts was widespread across the cases studied, and though they did not necessarily deal a death blow to the project, renegotiations always came at a further cost.

Telecommunications

Access to timely weather, market, and farming best practices information is no less important for agricultural development than access to transport infrastructure, regular and efficient irrigation, and energy. In Africa, as in much of the developing world, innovations in telecoms offer the potential to bring real change directly to the farm level long before more timely and costly investment in fixed infrastructure. Mobile phone penetration rates now exceed those of landlines and the industry is growing at an average annual rate of over 50% in the region. Mobile phone ownership in Africa increased from 54 million to almost 350 million between 2003 and 2008. Ownership rates under-represent actual usage, however, as many small vendors offer mobile access for calls or text messages. Even in rural areas, mobile penetration rates have now reached close to 42%. Mobile phones are becoming increasingly important tools in agricultural innovation, where they have been used to transfer and store money, check market prices and weather information, and even share farming best practices.

The case of India's e-Choupal (*choupal* is Hindi for a gathering place) illustrates the increasingly important role telecommunications can play in African agricultural innovation.[8] ITC is one of India's leading private companies, with annual revenues of US$2 billion. The company has initiated an e-Choupal effort that places computers with Internet access in rural farming villages; the e-Choupals serve as both a social gathering place for exchange of information and an electronic commerce hub. The e-Choupal system has catalyzed rural transformation that is helping to alleviate rural isolation, create more transparency for farmers, and improve their productivity and incomes. The system has also created a highly profitable distribution and product design channel for ITC—an e-commerce platform that is also a low-cost fulfillment system focused on the needs of rural India.

A computer, typically housed in a farmer's house, is linked to the Internet via phone lines or, increasingly, by a satellite connection, and serves an average of 600 farmers in 10 surrounding villages within about a five-kilometer radius. Each e-Choupal costs between US$3,000 and US$6,000 to set up and about US$100 per year to maintain. Using the system costs farmers nothing, but the host farmer, called a *sanchalak*, incurs some operating costs and is obligated by a public oath to serve the entire community; the *sanchalak* benefits from increased prestige and commissions for all e-Choupal transactions. The farmers can use the computer to access daily closing prices on local *mandi* (a government-sanctioned market area or yard where farmers sell their crops) as well as to track global price trends or find information about new farming techniques—either directly or, because many farmers are illiterate, via the *sanchalak*.

Farmers also use the e-Choupal to order seed, fertilizer, and other goods from ITC or its partners, at prices lower than those available from village traders; the *sanchalak* typically aggregates the village demand for these products and transmits the order to an ITC representative. At harvest time, ITC offers to buy the crop directly from any farmer at the previous day's closing price; the farmer then transports his crop to an ITC processing center where the crop is weighed electronically and assessed for quality. The farmer is then paid for the crop and a transport fee. "Bonus points," which are exchangeable for products that ITC sells, are given for above-normal quality crops.

Farmers benefit from more accurate weighing, faster processing time, prompt payment, and access to information that helps them to decide when, where, and at what price to sell. The system is also a channel for soil testing services and for educational efforts to help farmers improve crop quality. The total benefit to farmers includes lower prices for inputs and other goods, higher yields, and a sense of empowerment. At the same time, ITC benefits from net procurement costs that are about 2.5% lower and more direct control over the quality of what it buys. The system also provides direct access to the farmer and to information about conditions on the ground, improving planning and building relationships that increase its security of supply. The company reports that it recovers its equipment costs from an e-Choupal in the first year of operation and that the venture as a whole is profitable.

By 2010 there were 6,500 e-Choupals serving more than 4 million farmers in nearly 40,000 villages spread over 10 states. ITC is also exploring partnering with banks to offer farmers access to credit, insurance, and other services that are not currently offered or are prohibitively expensive. Moreover, farmers are beginning to suggest—and in some cases, demand—that ITC supply new products or services or expand into additional crops, such as onions and potatoes. Thus, farmers are becoming a source of product innovation for ITC.

By providing a more transparent process and empowering local people as key nodes in the system, the e-Choupal system increases trust and fairness. Improved efficiencies and potential for better crop quality make Indian agriculture more competitive. Despite the undependable phone and electrical power infrastructure that sometimes limit hours of use, the system also links farmers and their families to the world. Some *sanchalaks* track futures prices on the Chicago Board of Trade as well as local *mandi* prices, and village children have used the computers for schoolwork, games, and obtaining academic test results. The result is a significant step toward rural development.

The availability of weather information systems for farmers is also emerging as a critical resource. Although advances in irrigation infrastructure and technology are lowering farmers' dependency on weather, a second avenue to advance agricultural development is more accurate and more accessible weather information. To address the gap in accurate, timely, and accessible weather information in Africa, the Global Humanitarian Forum, Ericsson, World Meteorological Organization, Zain, and other mobile operators have developed a public-private partnership to (1) deploy up to 5,000 automatic weather stations in mobile network sites across Africa and (2) increase dissemination of weather information via mobile phones to users and communities—including remote farmers and fishermen.

Zain will host the weather equipment at mobile network sites being rolled out across Africa, as achieving the 5,000 target will require additional operator commitment and external financing. Mobile networks provide the necessary connectivity, power, and security to sustain the weather equipment. Through its Mobile Innovation Center in Africa, Ericsson will also develop mobile applications to help communicate weather information developed by national meteorological and hydrological services. Mobile operators will maintain the automatic weather stations and assist in the transmission of the data to national meteorological services. The initial deployment, already begun in Zain networks, focuses on the area around Lake Victoria in Kenya, Tanzania, and Uganda. The first 19 stations installed will double the weather monitoring capacity of the lake region.

Infrastructure and Innovation

One of the most neglected aspects of infrastructure investments is their role in stimulating technological innovation. Development of infrastructure in a country is often not enough to create sustained economic growth and lifestyle convergence toward that of developed countries. Technology learning is very important to a country's capacity to maintain current infrastructure and become competitive. In the first model of technology transfer, state-owned or privatized utility firms couple investment in public infrastructure with technological training programs, usually incorporated into a joint contract with international engineering firms. This type of capacity building lends itself to greater local participation in future infrastructure projects both in and out of the country.

The effectiveness of a comprehensive collaboration with foreign companies to facilitate both infrastructure building and technology transfer is seen in South Korea's contract with the Franco-British Consortium Alstom. The Korean government hoped to develop a high-speed train network to link Seoul with Pusan and Mokpo. The importance of the infrastructure itself was undeniable—the Korean Train Express (KTX) was meant to cross the country, going through a swath of land responsible for two-thirds of Korea's economic activity. In anticipation of the project, officials projected that by 2011, 120 million passengers would be using the KTX per year, leading to more balanced land development across South Korea.

However, while the project had the potential to increase economic activity and benefit the national industries in general, the project's benefits lay significantly in the opportunity "to train its workforce, penetrate a new industrial sector, and potentially take the lead in the high-speed train market in Asia."[9] In other words, Korea sought to obtain new technologies and the capacity to maintain and operate them. Under the contract with Alstom, which was finalized after 20 years of discussion, Alstom provided both the high-speed trains and railways that would help connect Seoul and Pusan and the training to help South Korea build and maintain its own trains.

From the beginning of negotiations, technology transfer was an important factor for the South Korean government. In 1992, bidding between Alstom, Siemens (a German group), and Mitsubishi (a Japanese group) commenced. After the bids were significantly slashed, Korean officials let it be known that in addition to price, financial structures and technology transfers would be major criteria during the final selection. It was in this category that Alstom successfully outbid the other consortia and won the contract, which specified that half of all production would occur in Korea, with 34 of the trains to be built by Korean firms. This would give Korea both production revenue and the experience of building high-speed trains—with the goal of one day exporting them. The contract also stipulated that 100% of Alstom's TGV (Train à Grande Vitesse) technology would be transferred to the 15 Korean companies that were to be involved in the project. Such technologies include industrial planning, design and development of production facilities, welding, manufacturing, assembly and testing carried out through operating and maintenance training, access to important documents and manuals for technical assistance, and maintenance supervision.

While the overall benefits of technology transfer are clear, more technologically advanced countries face some risks. One risk, known as the boomerang effect, affects the company that is transferring the technology—Alstom in this case. By giving the technological knowledge to South Korean companies, Alstom runs the risk of essentially creating its own competitor. This risk is especially high in this case because Alstom has transferred 100% of its TGV technology and 50% of the production to Korea. Low labor costs, weak contractual constraints, and Korea's known tendency to disregard intellectual property rights increase this risk. Other risks include unexpected shifts in economic conditions, currency devaluations, questionable competitive practices, hurried local production, lengthy and cumbersome administrative procedures, restrictive foreign payment rules, management weaknesses, and frail partnership involvement.

While these risks are indeed significant, they should not deter such agreements between countries. There are many ways to decrease such risks. For example, to make sure that payments are timely and that intellectual property rights are upheld, the company of interest should create a detailed contract with large penalties and disincentives for any violations.

Another step that should be taken is to maintain strong research and development projects to ensure that one's technology will always be superior. A good way to avoid the boomerang effect is to establish long-term relationships, such as Alstom established with Korea. A similar method of preventing the boomerang effect is to establish partnerships with local manufacturers. Finally, Alstom took strides to collaborate with established competitors, like the formation of EUROTRAIN with Siemens, to increase penetration into new markets.

And despite the numerous risks, training and technology transfer did not result in a loss for Alstom, for benefits included numerous cash payments, dividends, and income from giving access to its technology, selling equipment parts, and establishing separate ventures with Korean companies. Additionally, the project gave the company the opportunity to show the exportability of TGV to Asian markets. In particular, the reliability of Alstom's products and procedures was demonstrated in the partnership, making other countries more likely to work with the company. Experience in the Korean market, competitive advantages with respect to European countries, and new business opportunities were other advantages that increased Alstom's market share in Asia. Increased flexibility and experience with international markets as well as decentralized management also benefited Alstom. Finally, Alstom's technology became the international standard, leading to enormous competitive advantages for that company.

To facilitate the transfer, development, and construction of the high-speed railway system, the Korean government created the Korean High-Speed Rail Construction Authority (KHRC), whose mandate was to construct such systems at home and abroad, to research and find ways to improve the technology, and to oversee commercialization along the railway line. Issues with the project were soon revealed; after two tragedies—the collapse of the Songsu Bridge and of a large store in Seoul—Korean officials began to doubt its civil engineering capabilities. For this reason, KHRC decided to hire foreign engineers. After project delays and other issues, the last section of railway track from Taegu to Pusan was canceled and the building of 34 trains was postponed. However, after renegotiations, construction recommenced.

Other issues show the difficulty of such a project collaboration. A rift developed between France and South Korea due to a withdrawn agreement between two companies. Further, the TGV was unable to function in Korea during an unusually cold winter in 1996–1997, drawing questions and critiques from the Korean press. An economic crisis, which caused an abrupt depreciation of the Korean won against the U.S. dollar, made the purchase of goods and services from foreigners more expensive. A final rift was created by the election of President Kim Dae Jung in 1997, who was a vocal opponent of the KTX project. As seen here, exogenous interactions between the countries of interest can greatly affect the attempts to collaborate in technological transfer.

Despite the many risks of international technology transfer, the benefits far outweigh the costs. For example, by 2004, 100% of the TGV technology was transferred to Korea. Despite initial setbacks, ridership has increased greatly at the expense of other modes of transportation. More lines are expected to be built, and the success of the technology transfer has become apparent through the construction of the HSR-350x Korean-made train and the order of 19 KTX-II train sets in 2006 from Hyundai Rotem. It is claimed that these trains use 87% Korean technology. As for Alstom, their success is evident in the numerous contracts they have negotiated in the Asian markets.

In conclusion, the KTX project demonstrates how technology transfer can help developing countries to obtain advanced capabilities to build and develop infrastructure, leading to increased economic growth and productivity. The South Korean example serves as a model for African countries and applies to urban and rural projects alike. The lessons are particularly important considering the growing interest among African countries in investing in infrastructure projects.[10] While African countries will face their own unique issues, the KTX project illustrates costs and benefits that should be weighed in making such decisions and provides hope for new methods of technological dissemination. The tendency, however, is to view infrastructure projects largely in terms of their returns on investment and overall cost structure.[11] Their role in technological capacity building is rarely considered.[12] The growing propensity to want to leave infrastructure investments to the private sector may perpetuate the exclusion of public interest activities such as technological learning.[13]

One of the key aspects of the project was a decision by the government to set up the Korea Railroad Research Institute (KRRI). Founded in 1996, KRRI is the nation's principal railway research body. Its focus is improving the overall national railway system to maintain global competitiveness, with the goal of putting Korea among the top five leaders in railway technology. It works by bringing together experts from academia, industry, and government.

Regional Considerations

Roads, water facilities, airports, seaports, railways, telecommunications networks, and energy systems represent just a portion of the web of national and regional infrastructure necessary for food security, agricultural innovation, and agriculture-based economic development. Countries and regions must create comprehensive infrastructure investment strategies that recognize how each area is linked to the next, and investments must in many cases pool regional resources and cross numerous international borders. Transportation infrastructure is critical to move inputs to farms and products to market; widespread and efficient irrigation is essential for increasing yields and crop quality; energy is a vital input, particularly for value-added food processing; and telecoms are critical for the exchange of farming, market, and weather information. Alone, however, none of these investments will produce sustainable innovation or growth in agriculture. National and regional investment strategies will be needed to pool resources, share risks, and attract the private actors often critical to substantial investments in such ventures.

It seems obvious that roads would play a critical role in agricultural development, but they have often received inadequate investment. On-farm innovations are critical, but in many cases they depend on inputs that can only be delivered via roads, and they will be of very limited use if farmers have no way to reliably move their products to markets. Countries looking to improve their roads should carefully assess where their competitive advantages lie, identify which new or refurbished roads would best capitalize on those advantages, ensure that roads are placed within a broader plan for transportation infrastructure, and develop pre-construction plans for long-term maintenance.

Large roads and highways have garnered the bulk of capital and attention in much of the developing world, but smaller, lower-quality rural feeder roads often have significantly higher returns on investment—particularly in areas where major highways already exist. Learning from the Chinese experience, countries should carefully assess the relative return between larger highways and smaller rural-feeder roads, selecting the better investment.

National water policy and programs are notoriously Balkanized into fractious agencies and interest groups, often with competing objectives. This is a problem that countries across the world face, as is evidenced by the small American town of Charlottesville, Virginia. Charlottesville has no less than 13 separate water authorities representing its roughly 50,000 residents. As we saw with the positive example of Egypt (a country with significant water resource pressures but a highly advanced water management system), an initial step to success is streamlining government regulation of water issues under a single national agency, or family of agencies. Water policy and programs should be coordinated at the national, not state, level, and must also look across borders to neighboring states as many key issues in water, including power generation, agricultural diversion, and water quality, are often closely linked to key issues up- or downstream.

Many African states already face water shortages, and the threat of global climate change may further stress those limited resources. Bringing new water assets online through large irrigation projects is important, but those resources are limited; more economical use of water is just as if not more important. Central to this goal are farming techniques that get "more dollar per drip." As we saw earlier with the case of India, drip irrigation can be one solution. To overcome the initial capital hurdle, governments, companies, and banks could consider subsidizing and/or providing loans for the purchase of initial equipment.

As with water, energy issues often transcend national borders. In many cases, the best location to produce or sell power may be outside a country's borders. Regional cooperation will be essential for unlocking much of Africa's energy generation potential, as many projects will require far more investment than any one country can provide and involve assets that must span multiple national borders. To pool national resources and entice private capital for major energy products, regional organizations will need to help create strong, binding agreements to provide the necessary confidence not only to their member states but also to private companies and investors. The ECOWAS-led Western African Power Pool (WAPP) provides a good model for replication, but it is also an indicator of the high level of commitment and private capital that must be raised to push through large, regional power agreements.

Large power generation and transmission schemes are critical to agricultural development but in some cases may prove too lengthy, costly, or difficult to have large, timely impacts in remote rural areas. One way to complement these larger energy programs is to make additional investment in remote rural energy generation at the local or even farm level. Renewable technologies including solar, wind, biogas, bioethanol, and geothermal can be scaled for farms and small business and have the added advantage of requiring minimal transmission infrastructure and often a low carbon footprint. To encourage this production, governments could consider replicating Tanzania's Rural Energy Agency, which is funded by a small tax on sales from the national energy utility, as well as partnerships with NGOs, foundations, foreign governments, and businesses.

The transfer of knowledge is nearly as important to agricultural innovation as the transfer of physical inputs and farm outputs. Telecoms can play a unique role in the transfer of farming best practices as well as critical market and weather information. Most of Africa's telecom infrastructure is owned by the private sector. As we have seen from cases in India, China, and Africa, private companies can play a key role in the development of telecoms as a tool for agricultural innovation. Governments and regional bodies should work with major telecom providers and agribusinesses to form innovative partnerships that provide profits to companies and concrete benefits such as enhanced farming knowledge transfer and market and weather information.

Mobile phone penetration rates are growing faster in Africa than anywhere in the world. Mobile phones and the cell tower networks on which they depend provide a unique platform for the collection and even more important the dissemination of key information, including farming best practices, market prices, and weather forecasts. To reach scale, Africa's regional organizations should engage their member states, key telecom businesses, and NGOs to harness existing technologies such as SMS (and next generation technologies such as picture messaging and custom applications for mobiles) to provide farmers with access to key agricultural, market, and weather information.

Conclusion

Infrastructure investment is a critical aspect of stimulating innovation in agriculture. It is also one of the areas that can benefit from regional coordination. Indeed, the various RECs in Africa are already increasing their efforts to rationalize and coordinate infrastructure investments. One of the lessons learned from other countries is the importance of linking infrastructure investment (especially in key areas such as transportation, energy, water, and telecommunications) to specific agricultural programs. It has been shown that low-quality roads connecting farming communities to markets could contribute significantly to rural development. An additional aspect of infrastructure investment is the need to use such facilities as foundations for technological innovation. One strategic way to achieve this goal is to link technical training institutions and universities to large-scale infrastructure projects. The theme of education, especially higher technical training, is the subject of the next chapter.

5

HUMAN CAPACITY

Education and human capacity building in Africa have many well-publicized problems, including low enrollment and completion rates. One of the most distressing facts about many African school systems is that they often focus little on teaching students to maximize the opportunities that are available to them in their own communities; rather, they tend to prioritize a set of skills that is less applicable to village life and encourages children to aspire to join the waves of young people moving to urban areas. For some students, this leads to success, but for many more it leads to unfulfilled aspirations, dropout rates, and missed opportunities to learn crucial skills that will allow them to be more productive and have a better standard of living in their villages. It also results in nations passing over a chance to increase agricultural productivity, self-sufficiency, and human resources among their populations.

Education and Agriculture

African leaders have the unique opportunity to use the agricultural system as a driver for their economies and a source of pride and sustainability for their populations. About 36% of all African labor potential is used in subsistence agriculture. If that percentage of the population could have access to methods of improving their agricultural techniques, increasing production, and gaining the ability to transform agriculture into an income earning endeavor, African nations would benefit in terms of GDP, standard of living, infrastructure, and economic stability. One way to accomplish this is to develop systems—both formal and informal—to improve farmers' skills and abilities to create livelihoods out of agriculture rather than simply subsistence.

These systems start with formal schooling. Schools should include agriculture as a formal subject—from the earliest childhood experience to agricultural universities. They should consider agriculture an important area for investment and work to develop students' agricultural and technical knowledge at the primary and secondary levels. Universities should also consider agriculture an important research domain and devote staff and resources to developing new agricultural techniques that make sense for their populations and ecosystems. University research needs to stay connected to the farmers and their lifestyles to productively foster agricultural growth.

Decision makers should also look for ways to foster human capacity to make agricultural innovations outside of a traditional classroom. A variety of models incorporate this idea—from experiential and extension models to farmers' field schools, both discussed later in this chapter. Rural radio programs that reach out to farming communities and networks of farmers' associations spread new agricultural knowledge. In fact, there is a resurgence of radio as a powerful tool for communication.[1]

Governments and schools should treat agriculture as a skill to be learned, valued, and improved upon from early childhood through adult careers instead of as a last resort for people who cannot find the resources to move to a city and get an industrial job. Valuing the agricultural system and lifestyle and trying to improve it takes advantage of Africa's existing systems and capacities. In this way, many nations could provide significant benefits for their citizens, their economies, and their societies.

Nowhere is the missed opportunity to build human capacity more evident than in the case of women and agriculture in Africa. The majority of farmers in Africa are women. Women provide 70%–80% of the labor for food crops grown in Africa, an effort without which African citizens would not eat. Female farmers make up 48% of the African labor force. This work by women is a crucial effort in nations where the economy is usually based on agriculture.

Belying their importance to society and the economy, women have traditionally benefited from few of the structures designed to promote human capacity and ability to innovate. UNESCO estimates that only 45% of women in Africa are literate compared to 70% of men; 70% of African women do not complete primary school, and only about 1.5% of women achieve higher education. Of all of the disciplines, science and agriculture attract the fewest women.

For example, in Ghana, women account for only 13% of university-level agriculture students and 17% of scientists.[2] By not focusing on building the capacity of women, African states miss the chance to increase the productivity of a large portion of their labor force and food production workers. The lack of female involvement in education, especially science and agriculture, means there is an enormous opportunity to tap into skills and understandings of agricultural production that could help lead to more locally appropriate farming techniques and more thorough adoption of those techniques.

Gender and Agriculture

Although nearly 80% of agricultural producers in Africa are women, only 69% of female farmers receive visits from agricultural extension agents, compared to 97% of male farmers. Of agricultural extension agents, only 7% are female. In many places, it is either culturally inappropriate or simply uncommon for male extension agents to work with female farmers, so existing extension systems miss the majority of farmers. Additionally, as the Central American case below demonstrates, having extension workers who understand the experience of local farmers is central to promoting adoption. An important component of successful adoption is including female extension workers and educators in formal and informal settings.

Unequal distribution of education is the other critical factor in the misuse of women's contributions in agricultural production. Compared to the colonial period and the situation inherited at independence, considerable gains have been realized in general and female school attendance. Still, many countries have not yet achieved even universal primary school attendance. Gender inequality is most severe in contexts where general enrollment is lower.

Furthermore, several countries had a severe setback in the early 1980s, as their enrollment rates were either stagnant or declining. The persistent economic crisis meant that the previously agreed upon targets of universal primary enrollment in 1980 and then 2000 could not be met. Since the Dakar Conference of 2000 and creation of the Millennium Development Goals, new targets have been set for universal primary enrollment by 2015. However, at this stage, there is little doubt that most countries will not be able to reach this goal. By and large, countries with the lowest enrollment ratios from primary to higher education levels have the lowest enrollment ratios for their female populations.

African countries have shown considerable vitality in enrollment in higher education since the mid-1990s, following the lean years of the destructive structural adjustment programs. Nevertheless, African countries still have the lowest higher education enrollment in the world. Although there are a few exceptions in southern Africa (Lesotho is a unique case, where nearly three-fourths of the higher education students are females), in most African countries female enrollment is lower than that of males. Furthermore, the distribution of higher education students by discipline shows consistently lower patterns of female representation in science, technology, and engineering.

Considering African women's cultural heritage and continued central role in agriculture, it is a major paradox that their representation is so low in tracks where agricultural extension workers and other technicians and support staff and agricultural engineers are trained. Indeed, if there were any consistency between current educational systems and adequate human resource development, there would be at least gender parity in all the fields related to agriculture and trade. Yet only a few countries, such as Angola and Mozambique, have designed and implemented policies encouraging a high representation of females in science, including those fields related to agriculture. More generally, for both males and females, little effort is made in the educational system to promote interest in science in general and agriculture in particular.

It is vital to put more emphasis on involving women in agriculture and innovation, as well as helping female farmers build their capacity to increase productivity. There are several avenues to reach this goal. The first is women's training programs that focus specifically on agriculture. Another crucial avenue is emphasizing female participation in extension work—both as learners and extension agents, discussed later in this chapter.

The Uganda Rural Development and Training Program (URDT), which is establishing the African Rural University for women in the Kabala district of Uganda, is an innovative model of a program that focuses on building strong female leaders for careers in agriculture and on involving the community in every step of agricultural innovation. URDT students are fond of quoting the adage: "If you educate a man, you educate an individual. If you educate a woman, you educate a nation."

A key component of the program is to view individuals and communities holistically and to help people envision the future they want and to plan steps to get there. Their programming is tailored to locally identified needs that value the communities' lifestyles and traditions while allowing adoption of new technologies and improvement of production. URDT has had huge success in supporting change in the region since their founding as an extension project in 1987. Their impact has resulted in better food security, increased educational attainment, raised incomes for families across the district, better nutrition, and strong female leaders who engage in peace-building efforts and community improvements, among others. One driving factor is the innovative model of community-university interaction that focuses on women and agriculture.

URDT has a primary and secondary girls' school that focuses on developing girls' abilities in a variety of areas, including agricultural, business, and leadership skills, and encouraging them to bring their knowledge out to the community. At URDT Girls School, students engage in "Back Home" projects, where they spend some time among their families conducting a project that they have designed from the new skills they learned at school. Such projects include creating a community garden, building drying racks to preserve food in the dry season, or conducting hygiene education. Parents come to the school periodically to also engage in education and help the girls design the Back Home projects. School becomes both a learning experience and a productive endeavor; therefore, families are more willing to send children, including girls, to school because they see it as relevant to improving their lives.

URDT focuses on agriculture and on having a curriculum that is relevant for the communities' needs. They have an experimental farm where people can learn and help develop new agricultural techniques, as well as a Vocational Skills

Institute to work with local artisans, farmers, and businessmen who have not had access to traditional schooling. There is an innovative community radio program designed to share information with the broader community. URDT also runs an Appropriate and Applied Technology program that allows people from the community to interact with international experts and scientists to develop new methods and tools to improve their lives and agricultural productivity. A recent example is developing a motorcycle-drawn cart to help bring produce to market and improve availability and use of produce.

Governments can draw on this model to create effective learning institutions to support agriculture, and particularly women's and communities' involvement in it. The three key lessons of the model are to make sure that the school is working with and giving back to the community by focusing on its needs, which are often based around agriculture; to create a holistic program that sees how the community and the institution can work together on many interventions—technology, agriculture, market infrastructure, and education—to improve production and the standard of living; and to focus on women and girls as a driving force behind agriculture and community change, benefiting the whole society.

The crucial unifying factor is to integrate education at all levels, and the research processes of higher education in particular, back into the community. This allows the universities to produce technologies relevant to rural communities' needs and builds trust among the research, education, and farming communities.

Community-based Agricultural Education

Uganda is not alone in adopting this model. The government of Ghana established the University for Development Studies (UDS) in the northern region in 1992. The aim of the university is to bring academic work to support community development in northern Ghana (Brong-Ahafo, Northern, Upper East, and Upper West Regions). The university includes agricultural sciences; medicine and health sciences; applied sciences; integrated development studies; and interdisciplinary research. It relies on the resources available in the region.

UDS seeks to make tertiary education and research directly relevant to communities, especially in rural areas. It is the only university in Ghana required by law to break from tradition and become innovative in its mission. It is a multi-campus institution, located throughout northern Ghana—a region affected by serious population pressure and hence vulnerability to ecological degradation. The region is the poorest in Ghana, with a relatively high child malnutrition rate. The university's philosophy, therefore, is to promote the study of subjects that will help address human welfare improvement.

The pedagogical approach emphasizes practice-oriented, community-based, problem-solving, gender-sensitive, and interactive learning. It aims to address local socioeconomic imbalances through focused education, research, and service. The curricula stress community involvement and community dialogue, extension, and practical tools of inquiry.

Students are required to internalize the importance of local knowledge and to find effective ways of combining it with science. The curricula also include participatory rural appraisal, participatory technology development, and communication methodologies that seek to strengthen the involvement of the poor in development efforts.

An important component of the emphasis on addressing sustainable development is the third trimester practical field program. The university believes that the most feasible and sustainable way of tackling underdevelopment is to start with what the people already know and understand. This acknowledges the value of indigenous knowledge. The field program brings science to bear on indigenous knowledge from the outset.

Under this program, the third trimester of the academic calendar, lasting eight weeks, is exclusively for fieldwork. Students live and work in rural communities. Along with the people of the community, they identify development goals and opportunities and design ways of attaining them. The university coordinates with governmental agencies and NGOs in the communities for shared learning in the development process. The field exposure helps students build up ideas about development and helps them reach beyond theory. The impact of this innovative training approach is already apparent, with the majority of UDS graduates working in rural communities.

Early Agricultural Education

For children to engage in agriculture and understand it as a part of their life where they can build and develop skills and abilities to improve their future, it is necessary to continue their exposure to agricultural techniques and skills throughout their education. Equally important is the need to adapt the educational system to reflect changes in the agricultural sector.[3] Many rural African children will have been to the family farm or garden, and done some small work in the field, before they ever arrive at school. Children go with their mothers into the field from a very young age and so are likely to be familiar with local crops and the importance of the natural world and agriculture in their lives. Schools can capitalize on this early familiarity as a way to keep children engaged in the learning process and build on skills that will help them increase their production and improve their lives for the future.

School Gardens

One model to achieve early engagement is by having a school garden. Schools all over the world, from the United States and the United Kingdom, to Costa Rica and Ecuador, to South Africa and Kenya, use school gardens in various guises to educate their students about a set of life skills that goes beyond the classroom. School gardens come in many forms, from a plot of land in the school courtyard, to the children visiting and working in a broader community garden, to planting crops in a sack, a tire, or some other vessel. These gardens can use as many or as few resources as the community has to devote to them. The sack gardens especially require very few resources and can be cultivated in schools with little arable land and in urban areas. Students can also bring the sack garden model back home to their families to improve the family's income and nutrition.

Labor in the school garden should certainly not replace all other activities at the school, but as a complement to the other curriculum it can provide a place where students learn important skills and feel that they are productive members of their community. Children who participate in school gardens learn not only about growing plants, food, and trees—and the agricultural techniques that go along with this—but also about nutrition, food preparation, responsibility, teamwork, and leadership. As students get older, they can also use the garden and the produce it generates as a way to learn about marketing, economics, infrastructure needs, and organizing a business. Many schools have student associations sell their produce in local markets to learn about business and generate income.

School gardens have the added benefit of showing communities that the government recognizes agriculture as an important aspect of society and not as a secondary endeavor. Schools that provide education in gardening often overcome parents' reluctance to send children to school, as they teach a set of skills that the parents recognize as being important for the community—and parents do not see schooling as a loss of the child's potential labor at home. A government can increase this impact by involving the community in educational programs and curriculum decisions. Promoting buy-in from the community for the entire educational process encourages families to enroll more students and allows children to learn important skills.

Also, by valuing agriculture and enabling more productive work in the community, school gardens decrease the incentive for large migrations to urban areas. This also calms many parents' fears that a child who goes to school will leave home and not continue to work on the family farm. This emphasis on agriculture benefits both children and parents, by giving them access to a formal education and a way to increase agricultural productivity.

Semi-formal Schooling

Another model that can work to encourage children and young people to learn agriculture is a semi-formal schooling model. Here, children spend part of the day at school learning math, literacy, and traditional subjects, and part of the school day working in a field or garden. This second part of the day is a chance to generate some income for families as well as to learn new agricultural and marketing techniques. Generally, these kinds of programs are for older children who have never gone to primary school; they are taught in local languages instead of the official English or French of many formal school systems. This model can be adapted for adults as well to encourage literacy and the development and adoption of new agricultural techniques. In South Africa, this model is often referred to as a Junior Farmer's Field School, to get young people involved early in the experiential process of learning and creating new agricultural techniques.

School gardens, the inclusion of agriculture in the formal curriculum, and technical training models are all ways to promote children's experience with agriculture and help them develop the skills they need to improve their livelihoods into adulthood. These models place value on agriculture, the local community, and the process of experience to encourage children to learn new skills and engage in the natural world in a productive way.

Experiential Learning

There are several examples of how farmers can play a role in experimenting with new innovations, making them feel a sense of ownership of related tools and increasing the chance that other farmers will use the techniques. These examples also show how innovations work in the field and what changes are needed for better results.

Nonformal educational systems are crucial for reaching the population that is past the age for traditional primary schools and for encouraging local adoption of new techniques. Even if revolutionary new technologies exist at the research level, they can improve economies only if farmers use them, so getting information into the hands of local farmers, and especially women, is vital to the success of research endeavors and should be part of any plan for agricultural growth.

Two of the persistent obstacles to the adoption of new peanut varieties are the difficulty of obtaining the seeds and the reluctance to use new seeds without being sure how they will grow compared to the traditional variety. Farmers want many of the benefits that new seed varieties can bring—they typically prefer high-yield, high market value, pest-resistant, and high oil-content varieties—but often they cannot get the seeds or they are afraid the new seeds will fail. Without some guarantee that the new seeds will work, farmers are often unwilling to risk planting them, even if they are readily available, and these farmers are certainly unwilling to make the substantial investment of time and capital that is usually required to seek out and acquire new seed varieties. Not many rural farmers have the resources to go to the capital city and purchase experimental seed varieties from a research institute, and the risk of an unknown variety is often too high for a family to take.

One way of addressing this challenge is to give trial seed packages to pilot farmers or members of the local farmers' association to try on a portion of their land or on a test plot. This is a variation of the early adopters' model, which searches for members of a community who are willing and able to take some risk, and who then spread an idea to their peers. This strategy addresses both difficulties, since it allows for a trial with minimal risk, as well as a local source for new seed. Once the pilot farmer or association members grow the new variety of seed, they can sell it to their neighbors.

The International Crops Research Institute for the Semi-Arid Tropics (ICRISAT), in partnership with the Common Fund for Commodities has developed a trial package for new varieties of peanut in the Sahel (tested in Mali, Niger, and Nigeria) and disseminates it through pilot farmers and farmers' associations. These farming associations are often women's associations, since women traditionally need cash crops to be able to meet their families' economic needs but have even less access to improved seed varieties than men. In all of the countries, ICRISAT provided 17 kilograms of new seed varieties to their pilot farmers, as well as training in field management techniques that maximize the yield for their crop. The project's agents then asked local farmers to help distribute the new seed varieties through the members of their associations.[4]

Although the management techniques were imperfectly applied, and there is a cost associated with the new varieties and techniques, farmers using the new varieties experienced substantial returns on their investment. New varieties were 97% more profitable than traditional ones, so farmers earned almost twice as much on their investment as they would have with the types of seed already in use.

This is a story that has repeated itself often over the trials. ICRISAT has learned that the people most likely to adopt new peanut varieties—and who therefore make good pilot farmers—are those who are slightly younger, have smaller family sizes, and have relatively more access to resources, such as labor and land. These are the people who can afford to take a risk at the beginning, and when that risk pays off, they serve as a model for the other farmers in their community. They will then also serve as a source of local seed, which is very important, since farmers are most likely to use either their own seed stocks or stocks available from local markets.

Through this model, small investments can spread the use of modern seed varieties that have much higher yields and are more profitable to sell. These higher yields and profits ensure food security, including much-needed protein in rural diets, while improving the quality of life for the farmer.

As mentioned earlier, sending extension workers—either from governments or NGOs—into the field is a common practice that can be more or less effective, depending on who the extension agents are and how they handle the situation in the villages. Extension agents who are the peers of local villagers and practice the lessons they teach in a way that the other farmers can observe are usually the most effective.

Many countries in Africa have a variety of ethnic groups and regional subgroups who have different habits, speak different languages, and have different resources.

The further removed an extension agent is from the population with which he or she works—by barriers of language, socioeconomic status, gender, education, or tradition—the more difficult it is to convince people to adopt the technique. There is a tendency for people to decide that the idea is appropriate for someone like the extension agent, but not someone like themselves, even if they think that the idea is a good one. The comment, "That may be how that group does it, but it could never work in our village" is a common one for formal educators who come from a city or a different population subgroup. However, if the teacher is a peer, it is harder to make the distinction between their success and the potential success of each village farmer.

Governments can use the peer educator or farmer-to-farmer method to help spread information and new agricultural innovations across their entire rural population. By funding a few formal extension workers who train and help support a large network of peer educators, a government can reach most or all of the rural population, even if the groups are segregated by language, ethnicity, geography, or traditional farming techniques. Thus, a relatively small investment can have huge impacts on a country's agricultural processes and therefore on food security and the national economy.

Farmer Field Schools (FFS) provide a way for communities to test a new technique and adapt it to their own specific needs. Many agricultural technologies need to be adapted to local contexts once they leave the lab to ensure that they are practical for farmers and that people can adopt them into their current agricultural practices. The FFSs also allow for easier dissemination of new information because peers, as opposed to outsiders, are the teachers. This model also develops community ownership by encouraging local participation in new processes and leads to better adoption among participants.[5]

Local farmers participating in FFSs are often selected through local leadership structures or village farming associations. They plant one plot using the techniques that they currently use and a second plot with the new technology. At the end of the growing season, the farmers then come together to compare the costs, revenues, and profits between the old way and the new technology. In this way, farmers can see what works for them and can adapt the new method as seems appropriate during the growing season. Farmers also become invested in the process and have reason to believe that it will work for them.

Any organization—private or public—can start a Farmers' Field School. The resources needed are access to the new technology, be it a seed variety, a new fertilizer, or a new irrigation technique; a few extension agents to train a cadre of local farmers to spread the innovation; and a few follow-up visits to monitor the process and help villages interpret the results. These results should then move up to the national level to inform state policy and research. The following is an example of how an FFS can be used to address a specific problem.

Striga, often called witchweed, is a plant that grows in millet, sorghum, and other cereal fields across West Africa and causes a myriad of problems.[6] It can reduce crop yields between 5% and 80%, reduce soil fertility, and erode soil, all of which decrease the durability and profitability of rainy season–based agricultural systems. A single weed can produce more than 200,000 seeds, which remain viable in the soil for up to 10 years, making the plant very hard to eradicate. There are places where *Striga* infestation means that farmers lose money on every cereal crop they plant and are unable to feed their families or earn a living.

Nevertheless, there is a solution. In 2007, the International Crop Research Institute for the Semi-Arid Tropics (ICRISAT), the International Fund for Agricultural Development, and several European agricultural research organizations partnered with the Tominion Farmers' Union in Mali to implement a project that uses an integrated management system combining intercropping with beans or peanuts, reduced numbers of seeds per hole planted, and periodic weeding to control *Striga*. Using the FFS model, a few agricultural experts trained 75 local farmers to train their peers in integrated management techniques. These farmers then trained 300 others, and implemented the test plot procedure for their areas.

The results are impressive. *Striga* plants decreased, crop yields and profits increased, and many farmers decided to implement the process in their own fields. Farmers discovered that it was necessary for them to conduct three cycles of weeding rather than the two that the project originally recommended. This change has been formally adopted into the integrated management system. With these three weeding cycles, the incidence of *Striga* in the field went to zero in the test plots. Profits per hectare increased from $47 to $276, an improvement of nearly 500%. In some cases, villages went from a loss to a profit on their fields. The return on investment more than tripled, so while there was a slightly increased cost of the new methods, they more than paid for themselves. Many of the farmers involved in the 2007 study used the new methods in their own fields in 2008, and spread the message about the new techniques to their neighbors. This encouraged an enabling environment for the adoption of new technologies.

Another model is that of radio education—mentioned in the URDT case above—where extension education sessions are recorded in the appropriate local languages and played periodically on the radio. For many rural communities with low access to television and low literacy, this can be a crucial way to spread information to local farmers, especially if done in conjunction with another model, like the technical training or extension models that allow farmers to ask follow-up questions.

Innovation in Higher Agricultural Education

America's land-grant colleges pioneered agricultural growth by combining research, education, and extension services. The preeminent role of universities as vehicles of community development is reflected in the U.S. land-grant system.[7] The system not only played a key role in transforming rural America but also offered the world a new model for bringing knowledge to support community development. This model has found expression in a diversity of institutional innovations around the world. While the land-grant model is largely associated with agriculture, its adaptation to industry is less recognized. Universities such as the Massachusetts Institute of Technology (MIT) and parts of Stanford University owe their heritage to the land-grant system.[8] The drift of the land-grant model to other sectors is not limited to the United States. The central mission of using higher education to stimulate community development is practiced around the world in a variety of forms.

There are three models for entrepreneurial education in Brazil that have advanced to different stages of creating an "entrepreneurial university."[9] The Pontifical Catholic University of Rio de Janeiro, the Federal University of Itajubá, and the Federal University of Minas Gerais have all started to include entrepreneurship in the educational experience of their students. This experience often complements and coordinates with private sector initiatives, and in some cases companies fund parts of the curriculum. The interaction between academia, government, and industry allows for a broader approach and for a shifting of program goals.

The lessons from these three schools are that flexibility in curricula and the openness to partnering with other organizations—especially industry—allow universities to develop successful entrepreneurship programs that provide employment opportunities for their students as well as a chance to experience the culture of starting a business. The stimuli that lead universities to these activities might be an external change—lack of funding from the government—or an internal decision to shift focus. An institution that is more flexible, whose staff supports the change in a more unified way, is more likely to make the change toward becoming an "entrepreneurial university," which allows students to focus on not just having business know-how and the ability to work for or with large companies, but also on how to create jobs and opportunities for themselves and their peers. Universities must have the autonomy and the flexibility to adopt these programs as well as the ability to build networks with local actors. Ultimately, this will contribute to the nation's development.

African countries would be better served by looking critically at these variants and adapting them to their conditions. These institutional adaptations often experience opposition from advocates of incumbent university models. Arguments against the model tend to focus on the claim that universities that devote their time to practical work are not academic enough. As a result, a hierarchy exists that places such institutions either at the lower end of the academic ladder or simply dismisses them as vocational colleges.

The land-grant model is being reinvented around the world to address such challenges. One of the most pioneering examples in curriculum reform is EARTH University in Costa Rica, which stands out as one of the first sustainable development universities in the world.[10] It was created in 1990 through a US$100 million endowment provided by the U.S. Agency for International Development (USAID) and the Kellogg Foundation. Its curriculum is designed to match the realities of agribusiness.[11] The university dedicates itself to producing a new generation of agents of change who focus on creating enterprises rather than seeking jobs.

EARTH University emerged in a context that mirrors today's Africa: economic stagnation, high unemployment, ecological decay, and armed conflict. Inspired by the need for new attitudes and paradigms, EARTH University is a nonprofit, private, international university dedicated to sustainable agricultural education in the tropics. It was launched as a joint effort between the private and public sectors in the United States and Costa Rica. The Kellogg Foundation provided the original grant for a feasibility study at the request of a group of Costa Rican visionaries. Based on the study, USAID provided the initial funding for the institution. The original mission of the university was to train leaders to contribute to the sustainable development of the humid tropics and to build a prosperous and just society. Located in the Atlantic lowlands of Costa Rica, EARTH University admits about 110 students a year and has a total student population of about 400 from 24 countries (mainly in Latin America and the Caribbean) and faculty from 22 countries. Through its endowment, the university provides all students with 50% of the cost of tuition, room, and board.

In addition, the university provides scholarships to promising young people of limited resources from remote and marginalized regions. Nearly 80% of the students receive full or partial scholarship support. All students live on campus for four intensive years.

EARTH University has developed an innovative, learner-centered, and experiential academic program that includes direct interaction with the farming community.[12] Its educational process stresses the development of attitudes necessary for graduates to become effective agents of change. They learn to lead, identify with the community, care for the environment, and be entrepreneurial. They are committed to lifelong learning. There are four activities in particular within the curriculum that embodies EARTH University's experiential approach to learning.

Learning from Work Experience and Community Service

The first is the Work Experience activity, which is taken by all first-, second-, and third-year students and continues in the fourth year as the Professional Experience course. In the first and second years, students work in crop, animal, and forestry production modules on EARTH University's 3,300-hectare farm. In the first year, the work is largely a routine activity and the experience centers on the acquisition of basic skills, work habits, and general knowledge and familiarity with production. In the second year, the focus changes to management strategies for these same activities. Work Experience is later replaced with Professional Experience. In this course students identify work sites or activities on campus that correspond with their career goals. Students are responsible for contacting the supervisors of the campus operations, requesting an interview, and soliciting "employment." Upon agreement, supervisors and students develop a joint work plan that the student implements, dedicating a minimum of 10 hours per week to the "job." The second activity is an extension of the Work Experience course. Here third-year students work on an individual basis with small, local producers on their farms. They also come together in small groups under the Community Outreach program that is integral to the learning system. Community outreach is used to develop critical professional skills in students, while at the same time helping to improve the quality of life in nearby rural communities. The third-year internship program emphasizes experiential learning. The 15-week internship is required for all students in the third trimester of their third year of study. It is an opportunity for them to put into practice all they have learned during their first three years of study. For many of them it is also a chance to make connections that may lead to employment after graduation. The international character of the institution allows many students the opportunity to follow their interests, even when they lead to internship destinations other than in their home country.

Sharpening Entrepreneurial Skills

The fourth activity is the Entrepreneurial Projects Program. EARTH University's program promotes the participation of its graduates in the private sector as a critical means by which the institution can achieve its mission of contributing to the sustainable development of the tropics. The development of small and medium-sized enterprises (SMEs) is a powerful way to create new employment and improve income distribution in rural communities. For this reason, the university stresses the development of an entrepreneurial spirit and skills. Courses in business administration and economics combined with practical experience prepare the students to engage in business ventures upon graduation.

This course provides students the opportunity to develop a business venture from beginning to end during their first three years at EARTH University. Small groups of four to six students from different countries decide on a relevant business activity. They conduct feasibility studies (using financial, social, and environmental criteria), borrow money from the university, and implement the venture. This includes marketing and selling the final product. After repaying their loan, with interest, the group shares the profits. This entrepreneurial focus has permeated all aspects of the university's operations and prepared students to become job creators and agents of change rather than job seekers. About 17% of the university's 1,100 graduates run their own businesses. The university also manages its own profitable agribusiness, which has resulted in strong relationships with the private sector. When the university acquired its campus, it decided to continue operating the commercial banana farm located on the property. Upon taking over the farm, the university implemented a series of measures designed to promote more environmentally sound and socially responsible production approaches.

Going Global

EARTH University has internationalized its operations. It signed an agreement with U.S.-based Whole Foods Market to be the sole distributor of bananas in their stores. The university also sells other agricultural products to the U.S. market. This helps to generate new income for the university and for small farmers while providing an invaluable educational opportunity for the students and faculty. In addition to internships, students have access to venture capital upon graduation. The university uses part of the income to fund sustainable and organic banana and pineapple production research. Over the years the university has worked closely with African institutions and leaders to share its experiences. Following nearly seven years of study through workshops, discussions, training courses, and site visits, African participants agreed to the importance of reforms in their own university systems, especially through the creation of new universities along the lines of the EARTH model. The case of EARTH University is one of many examples around the world involving major collaborative efforts between the United States and east African countries to bring scientific and technical knowledge to improve welfare through institutional innovations. Such experiences, and those of U.S. land-grant universities, offer a rich fund of knowledge that should be harnessed for Africa's agricultural development and economic growth.

Such models show how to focus agricultural training as a way to improve practical farming activities. Ministries of sustainable agriculture and farming enterprises in east African countries should be encouraged to create entrepreneurial universities, polytechnics, and high schools that address agricultural challenges. Such colleges could link up with counterparts in developed or emerging economies as well as institutions providing venture capital and start to serve as incubators of rural enterprises. Establishing such colleges will require reforming the curriculum, improving pedagogy, and granting greater management autonomy. They should be guided by the curiosity, creativity, and risk-taking inclination of farmers.

Policy Lessons

The challenges facing African agriculture will require fundamental changes in the way universities train their students. It is notable that most African universities do not specifically train agriculture students to work on farms in the same way medical schools train students to work in hospitals. Part of the problem arises from the traditional separation between research and teaching, with research carried out in national research institutes and teaching in universities. There is little connection between the two in many African countries. This needs to radically change so that agricultural education can contribute directly to the agricultural sector.

A number of critical measures are needed at the regional and national levels to achieve this goal.[13] The first should be to rationalize existing agricultural institutions by designating some universities as hubs in key agricultural clusters. For example, universities located in proximity to coffee production sites should develop expertise in the entire value chain of the coffee industry. This could be applied to other crops as well as livestock and fisheries. Such universities could be designed around existing national research institutes that would acquire a training function as part of a regional rationalization effort. Such dedicated universities would not have monopoly over specific crops but should serve as opportunities for learning how to connect higher education to the productive sector.

Internally, the universities should redefine their academic foci to adjust to the changes facing Africa. This can be better done through continuous interaction among universities, farmers, businesses, government, and civil society organizations. Governance systems that allow for such continuous feedback to universities will need to be put in place.

The reform process will need to include specific measures. First, universities need a clear vision and strategic planning for training future agricultural leaders with a focus on practical applications.[14] Such plans should include comprehensive road maps on how to best recruit, retain, and prepare future graduates. These students should be prepared in partnership with key stakeholders.

Second, universities need to improve their curricula to make them relevant to the communities in which they are located. More important, they should serve as critical hubs in local innovation systems or clusters. The community focus, however, will not automatically result in local benefits without committed leadership and linkages with local sources of funding.[15] The recent decision by Moi University in western Kenya to acquire an abandoned textile mill and revive it for teaching purposes is an example of such an opportunity.

Third, universities should give students more opportunities to gain experience outside the classroom. This can be done through traditional internships and research activities. But the teaching method could also be adjusted so that it is experiential and capable of imparting direct skills. More important, such training should also include the acquisition of entrepreneurial skills.

Fourth, continuous faculty training and research are critical for maintaining high academic standards. Universities should invest more in undergraduate agricultural educators to promote effective research and teaching and to design new courses.

Finally, it is important to establish partnerships among various institutions to support and develop joint programs. These partnerships should pursue horizontal relationships and open networking to generate more synergy and collaboration, encourage sharing of resources, and foster the exchange of students and faculty. This can be done through regional exchanges that involve the sharing of research facilities and other infrastructure.

Providing tangible rewards and incentives to teachers for exemplary teaching raises the profile of teaching and improves education. In addition, establishing closer connections and mutually beneficial relationships among all stakeholders (academia and industry, including private and public institutions, companies, and sectors) should generate further opportunities for everyone.

Lifelong Learning through the Private Sector

The roles of the private and public sector in lifelong learning opportunities are illustrated by the case of Peru's relatively high-tech asparagus industry.[16] Both public and private programs offer industry-specific training for employees and build on the skills that many workers get from experience in the formal education sector. Those working at the managerial level tend to receive training from La Molina—the national agricultural university. There is a tension between private and public sector training, as hiring managers tend to perceive graduates of private education as being of higher quality, although the public sector is able to produce more graduates and therefore better meet the industry demand for workers. Ultimately, the best arrangement is some combination of public and private sector education and training, as Peru has high secondary and tertiary school enrollment compared to many other Latin American countries.

Asparagus exporting requires a high level of skill because of the need to keep the asparagus under controlled conditions and package it in appropriate weights. The success of this industry relies upon investment in long-term learning for employees. There is a great emphasis on on-the-job training, whereby employees learn a specific set of job-related skills. In addition, there are both private and public vocational training programs for adults. Employers give consistently better reviews to those workers who receive on-the-job training or who complete the private sector training programs than to those who graduate from government-run programs. Students are willing to pay for private training because the curriculum and schedule are more flexible, and they allow the students to continue their employment, in contrast to the more rigid structure in public institutions. These private institutions also generally include an internship—which serves as both student training and a relatively low-cost way for employers to recruit skilled students.

Nevertheless, these programs are not without problems; they produce fewer students than the industry needs, and they rely on employees having at least primary or some secondary education, largely as a result of Peru's relatively high levels of enrollment in secondary schools.

A high proportion of managers graduate from La Molina with degrees in agronomy or engineering. La Molina not only trains many of the skilled workers in management and agronomic skills, but it also conducts much of the research that the industry uses to have its crops meet international export standards. La Molina also conducts technology transfer with countries like the United States and Israel to adapt new techniques to local realities.

There are several training models to help farmers and plant workers acquire the skills they need. First are the private models of on-the-job training, which range from informal mentoring in the first two weeks of work to Frio Aéreo's (a consortium of 10 partners that is concerned with managing the cold chain) formalized internship program and weekly training sessions during the slower seasons. Second are private universities such as Universidad Privada Antenor Orrego that train technicians and managers; there are also public institutions with similar goals, whose graduates tend to be less valued by employers. Additionally, there is a public sector youth training program that aims to help young graduates become successful agricultural entrepreneurs. Finally, there is El Centro de Transferencia de Tecnologia a Universitarios, which uses holistic approaches to develop agricultural entrepreneurs, either by giving students plots of land that they must pay for over several years, or working with small farmers who already own land. The model that works with smallholders requires an investment of about US$33,000 per farmer.

Private sector initiatives have so far been more successful at training older workers in the necessary techniques. However, much of the system depends on workers having initial basic public education, as well as the managerial expertise and public goods that La Molina provides. Successful training for high-skill industries requires a combination of private sector initiatives and a solid foundation of public education and research.

Conclusion

The current gaps in educational achievement and the lack of infrastructure in many African school systems are an opportunity for governments to adopt more community-driven models that prioritize education in a holistic way that improves community involvement, child achievement, agricultural production, and the standard of living for rural populations. Acknowledging that agriculture is both a valued traditional lifestyle and a huge potential driver of economic growth, and changing educational programming to respect these goals, will go a long way toward encouraging basic education and improving people's lives.

No new agricultural technology, however cutting-edge and effective, can improve the situation if people are unable to access it and use it. Farmers need to have the capacity to adopt and understand new technologies, and the system needs to develop to meet their needs and to enable them. Since most of the farmers in Africa are women, an important component of these systems will be including women in all parts of the process: education, capacity building, and technology innovation.

6

ENTREPRENEURSHIP

The creation of agricultural enterprises represents one of the most effective ways to stimulate rural development. This chapter will review the efficacy of the policy tools used to promote agricultural enterprises, with a particular focus on the positive, transformative role that can be played by the private sector. Inspired by such examples, this chapter will end by exploring ways in which African countries, subregional, and regional bodies can create incentives that stimulate entrepreneurship in the agricultural sector. The chapter will take into account new tools such as information and communication technologies and the extent to which they can be harnessed to promote entrepreneurship.

Agribusiness and development

Economic change entails the transformation of knowledge into goods and services through business enterprises. In this respect, creating links between knowledge and business development is the most important challenge facing agricultural renewal in east African countries. The development of small and medium-sized enterprises (SMEs) has been an integral part of the development of all industrialized economies. This holds true in Africa. Building these enterprises requires development of pools of capital for investment; of local operational, repair, and maintenance expertise; and of a regulatory environment that allows small businesses to flourish. Africa must review its incentive structures to promote these objectives.[1]

A range of government policy structures is suitable for creating and sustaining enterprises–from taxation regimes and market-based instruments to consumption policies and changes in the national system of innovation. Policy makers also need to ensure that educational systems provide adequate technical training. They need to support agribusiness and technology incubators, export processing zones, and production networks as well as sharpen the associated skills through agribusiness education.

Banks and financial institutions also play key roles in fostering technological innovation and supporting investment in homegrown domestic businesses. Unfortunately, their record in promoting technological innovation in Africa has been poor. Capital markets have played a critical role in creating SMEs in other developed countries. Venture capitalists not only bring money to the table; they also help groom small and medium-sized start-ups into successful enterprises. Venture capital in Africa, however, barely exists outside of South Africa and needs to be introduced and nurtured.

Much of the effort to promote venture capital in developing countries has been associated with public sector initiatives whose overall impact is questionable.[2] One of the possible explanations for the high rate of failure is that many of these initiatives are not linked to larger strategies to create local innovation systems. Venture capital is only one enabling species in a complex innovation ecosystem.[3] It does not exist in an institutional or geographical vacuum and appears to obey the same evolutionary laws as other aspects of innovation systems.[4] It is therefore important to look at examples of geographical, technological, and market aspects of venture capital. The legal elements needed to create institutions are only a minor part of the challenge.

One critical starting point is "knowledge prospecting," which involves identifying existing technologies and using them to create new businesses. African countries have so far been too isolated to benefit from the global stock of technical knowledge. They need to make a concerted effort to leverage expertise among their nationals residing in other countries. Such diasporas can serve as links to existing know-how, establish links to global markets, train local workers to perform new tasks, and organize the production process to produce and market more knowledge-intensive, higher value added agricultural products.

Advances in communications technologies and the advent of lower-cost high-speed Internet will also reduce this isolation dramatically. The laying of new fiber-optic cables along the coasts of Africa and, potentially, the use of lower-latency satellite technology can significantly reduce the price of international connectivity and will enable African universities and research institutions to play new roles in rural development. The further development of Internet exchange points (ISPs) in east Africa where they do not currently exist is also important. ISPs enable Internet traffic to be exchanged locally, rather than transverse networks located outside the continent, improving the experience of users and lowering the cost to provide service.

Much is already known about how to support business development. The available policy tools include direct financing via matching grants, taxation policies, government or public procurement policies, advance purchase arrangements, and prizes to recognize creativity and innovation. These can be complemented by simple ways to promote rural innovation that involve low levels of funding, higher local commitments and consistent long-term government policy.

For example, China's mission-oriented "Spark Program," created to popularize modern technology in rural areas, had spread to more than 90% of the country's counties by 2005. The program helped to improve the capability of young rural people by upgrading their technological skills, creating a nationwide network for distance learning, and encouraging rural enterprises to become internationally competitive. The program was sponsored by the Minister of Science and Technology.[5]

There is growing evidence that the Chinese economic miracle is a consequence of the rural entrepreneurship that started in the 1980s. This contradicts classical interpretations that focus on state-led enterprises as well as receptiveness to foreign direct investment. The creation of millions of township and village enterprises (TVEs) in provinces such as Zhejiang, Anhui, and Hunan played a key role in stimulating rural industrialization.[6]

Over the past 60 years, China has experimented extensively with policies and programs to encourage the growth of rural enterprises that provide isolated agricultural areas with key producer inputs and access to post-harvest, value-added food processing. Despite the troubled early history, by 1995 China's TVEs had helped bring about a revolution in Chinese agriculture and had evolved to account for approximately 25% of China's GDP, 66% of all rural economic output, and more than 33% of China's total export earnings.[7]

Most of the TVEs have become private enterprises and focus in areas outside agricultural inputs or food processing. Agricultural support from TVEs remains relevant, however, particularly as a model by which other countries may be able to increase farmers' access to key inputs such as fertilizers and equipment, as well as value-added processing of raw agricultural products.

With few rural-urban connecting roads and weak distribution systems, the Chinese government moved to resolve these agricultural input and post-harvest processing constraints by creating new enterprises in rural areas. China's initial rural enterprise strategy therefore focused on the so-called five small industries that it deemed crucial to agricultural growth: chemical fertilizer, cement, energy, iron and steel, and farm machinery. With strong backward linkages between these rural enterprises and Chinese farmers, agricultural development in China grew substantially in the late 1970s and 1980s through farmland capital construction, chemical fertilization, and mechanization.

This expansion in agricultural productivity, coupled with high population growth, led to a surplus of labor and a scarcity of farmland. As a consequence, China's rural enterprises increasingly shifted from supplying agricultural producer inputs to labor-intensive consumer goods, for domestic and (after 1984 market reforms) international markets. From the mid-1980s to the 1990s, China's TVEs saw explosive growth in these areas while they continued to supply agricultural producers with access to key inputs, new technologies, and food processing services. In 1993, 8.1% of total TVE economic output came from food processing, while chemicals (including fertilizer) accounted for 10%, building materials 12%, and equipment (including for farms) 18%.

The most successful TVEs were those with strong links to urban and peri-urban industries with which they could form joint ventures and share technical information; those in private ownership; and those with a willingness to shift from supplying producer inputs for farmers to manufacturing consumer goods for both domestic and international markets.

China's experience with rural enterprises confirms that they may provide a mechanism through which developing states can enhance rural access to key agricultural inputs such as fertilizers and mechanization, as well as value-added post-harvest food processing. Rural enterprises may make the most sense in areas where farm-to-market roads cannot be easily established to achieve similar backward and forward linkages. In addition to sparking agricultural productivity and growth, rural enterprises may also help provide employment for farm laborers displaced by agricultural mechanization. By keeping workers and economic activity in rural areas, China has helped expand rural markets, limit rural-urban migration, and create conditions under which it is easier for the government to provide key social services such as health care and education.

Despite the fact that TVEs enjoyed government support through financing and technical assistance, they enjoyed a degree of autonomy in their operations. The emergence of rural markets in China not only contributed to prosperity in agricultural communities, but it also provided the impetus for the modernization of the economy as a whole.[8] Furthermore, the TVEs also became a foundation for creating entrepreneurial leadership and building managerial and organizational capacity.[9]

Such entrepreneurial initiatives will succeed in the absence of consistent and long-term policy guidance on the one hand, and autonomy of action on the part of farmers and entrepreneurs, on the other hand. The latter is particularly critical because a large part of economic growth entails experimentation and learning. Neither of these can take place unless farmers and associated entrepreneurs have sufficient freedom to act. In other words, development has to be viewed as an expression of human potentialities and not a product of external interventions.

The Seed Industry

The seed sector in sub-Saharan Africa is dominated by informal supply systems with farm-saved seeds accounting for approximately 80% of planted seeds. Improving smallholder farmers' access to new high-yielding varieties and hybrid crops requires better coordinated marketing efforts and expanded distribution systems. Because of their small size and market orientation, small to medium-sized emerging seed companies have a potential competitive advantage in meeting the needs of smallholder farmers. Emerging seed companies—the nexus of publicly supported agricultural biotechnology and newly created market opportunities for the private sector—can promote food security and welfare improvement within economically disadvantaged rural communities.[10]

However, these emerging domestic companies have limited financial and managerial resources and are often hampered by complex and bureaucratic legal frameworks. As infants in the industry, small to medium-sized domestic seed companies need short-term assistance, especially in establishing a solid financial base and developing management capacity.

Maize is a staple in southern and eastern Africa, yet the amount of produce and the acreage of maize have not increased much over the years, even though the number of grain producers has quadrupled. However, the seed sector faces major challenges. Although less monopolized now, the seed sector in a majority of African countries is far from being efficient. The seed industry suffers from five levels of bottlenecks, producing an adverse effect on the maize seed value chain across the region. The first bottleneck is government political and technical policies. Import procedures, for instance, are cumbersome enough in Tanzania to dissuade seed import while in Zimbabwe, during the economic crisis, the government banned seed exports.[11]

Second, establishing a seed company has a high initial cost, requiring access to credit; the company also needs qualified manpower. Third, the production of seed suffers from a lack of adequate and adapted input, from expensive production costs and lack of production credit, and from poor weather and unfavorable land policies. Fourth, poor infrastructure in the value chain, such as poor retail networks or sales points, jeopardize marketing and access to the farmers. Last, farmers tend to have low demand for seeds.

Millet and Sorghum Production in India

Millions of small-scale farmers in India live in harsh environments where rainfall is limited and irrigation and fertilizer are unavailable.[12] In these harsh areas, many farmers have long grown sorghum and pearl millet—hardy crops that can thrive in almost any soil and survive under relatively tough conditions. Production from these crops was low, however, and so were returns to farmers, until improved, higher-producing varieties were developed and distributed starting in the 1970s. Since then, a succession of more productive and disease-resistant varieties has raised farmers' yields and improved the livelihoods of about six million millet-growing households and three million sorghum-growing households. Although public funding was the key to developing this improved genetic material, it has been private seed companies that have helped ensure that these gains were spread to, and realized by, the maximum number of Indian farmers.

The success and sustainability of these improved varieties resulted from interventions by the Indian government and the international community, as well as the increasing inclusion of private industry and market-based solutions in seed sale and distribution. Three key interventions include increased investments in crop improvements during the 1970s; the development of efficient seed systems with a gradual inclusion of the private sector in the 1980s; and the liberalization of the Indian seed industry in the late 1990s. By allowing farmers to grow the same amount of millet or sorghum using half as much land, these improved varieties have made it possible for farmers to shift farmland to valuable cash crops and thereby raise their incomes. Our analysis will focus on the key role played by the last of these three innovations, the establishment of a private seed industry.

Government Investment in Research

The first advances in millet and sorghum research in India resulted from the efforts of a range of government institutions. These programs organized government research and in many locations tested for improved characteristics of hybrids and varieties—through state agricultural universities, research institutes, and experiment stations. Joint efforts by these institutions resulted in the release of a succession of pearl millet hybrids offering yield advantages. Since the mid-1960s, average grain yields have nearly doubled, even though much of the production of millet has shifted to more marginal production environments. Production of pearl millet in India currently stands at nine million tons, and hybrids are grown in more than half of the total national pearl millet area of 10 million hectares.

Cultivating the Seed Industry

At the beginning of the Green Revolution, the Indian government and key state governments decided that state extension services and emerging private seed companies could not distribute enough seed to allow for the large-scale adoption of new varieties. The government decided to create state seed corporations, the first of which evolved out of the G. B. Pant University of Agriculture and Technology in Pantnagar. This corporation then became a model for the National Seed Corporation and other state seed corporations.

The Indian government, with the financial support of the World Bank and technical assistance from the Rockefeller Foundation, financed the development of state seed corporations in most major Indian states in the 1960s. Gradually, these state seed corporations replaced state departments of seed production and formed the nascent foundations of a formal seed industry. Often, formal seed industries are taken for granted, especially in industrial countries, where agriculture is extremely productive. But in India, as in many other countries, seed industries are still emerging. The problem stems from the limited profitability of seeds. When farmers are able to plant and save seeds from one season to the next without losing much in terms of yield and output, there is little need for them to purchase new seeds—and little opportunity for seed producers to sell new seeds.

It is only when commercial seeds offer clear advantages in terms of quality and performance that farmers become more willing to purchase them. When improvements are bred into a crop, for example, farmers must buy or otherwise gain access to the improved seed to realize the benefits of breeding. Farmers must also buy seeds to realize the full benefits of hybrids, the yields of which tend to drop when grain from harvests is saved and planted in the next season. But seed industries do not emerge simply by themselves. The right rules and regulations must be in place to encourage private investment in the industry and to limit the role of the public sector where it is a less-efficient purveyor of seed to farmers. In India, this institutional framework for the development of a seed industry emerged with the Indian Seed Act in 1966. The nascent Indian seed industry was heavily regulated under the act, however, with limited entry and formation of large private firms—domestic or foreign. Private seed imports for both commercial and research purposes were restricted or banned, ostensibly to protect smallholders from predatory corporate practices.

Emergence of the Private Seed Industry

Since the 1970s, the private sector has played an ever-increasing role in developing improved varieties of millet and sorghum and distributing them to farmers through innovative partnerships with public sector agencies. In 1971, India began deregulating the seed sector, relaxing restrictions on seed imports and private firms' entry into the seed market. This change, combined with a new seed policy in 1988, spurred enormous growth in private sector seed supplies in India. Currently, the Indian market for agricultural seed is one of the biggest in the world.

Sorghum and pearl millet breeding by private companies began around 1970, when four companies had their own sorghum and pearl millet breeding programs. By 1985 this number had grown to 10 companies. In 1981, a private company developed and released the first hybrid pearl millet. One major reason for the spurt in private sector growth was the strong public sector research on sorghum and millet. International agricultural research centers such as the International Crops Research Institute for the Semi-Arid Tropics (ICRISAT) exchanged breeding material with public and private research institutions. National agricultural research centers such as the Indian Council of Agricultural Research (ICAR) and agricultural universities provided breeder seed not only to the national and state seed corporations but also to private seed companies to be multiplied and distributed through their company outlets, farmer cooperatives, and private dealers. For private firms, public institutions like ICRISAT, ICAR, and state universities provided invaluable genetic materials, essentially free of charge.

Today, more than 60 private seed companies supply improved pearl millet to small-scale farmers and account for 82% of the total seed supply, while more than 40 companies supply improved sorghum, accounting for 75 % of supply. Many of these companies benefit not only from the availability of public research on improved pearl millet and sorghum but also from innovative partnerships that specifically aim to disseminate new materials to the private sector. The most recognized of these partnerships is ICRISAT's hybrid consortia, developed in 2000–01. Private companies pay a membership fee to ICRISAT to receive nonexclusive access to hybrid parent lines that they can then use for the development and marketing of their own seed products. Although no single company has a monopoly over an individual line—all companies can use them for their own purposes as they choose—the market is currently large enough to allow all companies to compete for the small-holders' business.

The ultimate beneficiaries of this public-private system are the millions of small-scale farmers who grow sorghum and millet. Public research agencies contribute genetic materials and scientific expertise to improve crop varieties when the incentives for private sector involvement are limited. Then, private companies take on the final development of new vari-eties and seed distribution—tasks to which they are often better suited than are public agencies. In this way, the benefits of crop improvements are delivered directly to farmers, who find them worthwhile enough to support financially.

All three elements of the Indian intervention to improve sorghum and pearl millet hybrids were important. First, the investments in public sector plant-breeding and crop-management research were made by the national government, state governments, and international agricultural research centers. When hybrids of sorghum and millet were first being developed, all three of these groups contributed genetic material that benefited farmers directly and provided the basis for private researchers to develop new varieties. Second, the government invested in seed production in public and private institutions. The Indian government and state govern-ments, with the help of donors, made major investments in government seed corporations that multiplied the seeds of not only wheat, rice, and maize, but also pearl millet and sorghum. Seed laws were written and enforced to allow small private sector seed companies to enter the seed business and make profits. The government also provided training for people involved in the seed industry in both public and private institutions.

Third, and most important, India liberalized the seed sector starting in the mid-1980s. Instead of allowing state seed corporations to become regional monopolies, the government opened the doors to investment by large Indian firms and allowed foreign direct investment in the sector. This change, coupled with continuing investments in public plant breeding and public-private partnerships, has continued to provide private firms with a steady stream of genetic materials for developing proprietary hybrids. India also benefits from a seed law that allows companies to sell truthfully labeled seed without having to go through costly and time-consuming certification and registration processes for new hybrids and varieties. The result is a vibrant and sustainable supply of seed of new cultivars that are drought tolerant and resistant to many pests and diseases.

Africa's Seeds of Development Program

The Seeds of Development Program (SODP) is designed to improve access to appropriate, good quality, and competitively priced crop seeds for low-income smallholder farmers in east and southern Africa. This has been achieved by focused management training for over 30 small to medium-sized local seed companies in the region such as Vitoria Seeds in Uganda, Freshco Seeds in Kenya, Kamano Seeds in Zambia, Qualitá in Mozambique, and Seed Tech in Malawi. Utilizing a grant from the UK Department for International Development (DFID), which was issued in 2006, SODP aims to achieve the purpose through two main outputs. The first output is to increase the scale of the program by enrolling additional participants from countries already involved in SODP and, in addition, bringing new countries into the program. The second output is to increase the scope of the program through widening program activities.[13]

SODP management training is showing results, with SODP companies selling seed around 20% cheaper than their larger competitors. SODP networking is also providing new ways of doing business and opportunities for partnerships across countries. The link with the Alliance for a Green Revolution in Africa has proved useful for some of the companies. In the proposed follow-up SODP program, a major focus of development will be facilitating effective alliances between the two main SODP fellow categories (full service seed companies and seed traders) to enable each to exploit their niche within the smallholder market for seed.

SODP is an innovative program, valued by its participants. Performance indicators are impressive—maize seed sales up by 54% between 2006 and 2007; full-time employment up by 19%; and sales revenue up by 35%. Company sales data also show that the bulk of sales (more than 80%) go to smallholder farmers. By offering a wider variety of seeds, including higher-yielding, disease- and drought-resistant varieties, and other inputs such as fertilizers, SODP companies help smallholder farmers increase food security for their families and communities. The evidence available shows that SODP members are producing and selling seed to smallholders at significantly lower prices than their larger-scale competitors.

Food Processing

Transformations in the food processing sectors of developing countries are increasingly seen as strategic from the point of view of export earnings, domestic industry restructuring, and citizens' nutrition and food security.[14] The widespread adoption by developing countries of export-led growth strategies has drawn attention to the economic potential of their food processing sectors, particularly in the light of the difficulties faced by many traditional primary commodity export markets. Food processing can be understood as post-harvest activities that add value to the agricultural product prior to marketing. In addition to the primary processing of food ingredients, it includes, therefore, final food production on the one hand and the preparation and packaging of fresh products. To better understand the role of food processing in African agricultural development, this chapter will examine several cases of successful African food processing start-ups as well as the role new technologies (particularly radio and video) can play in teaching farmers how to add value with post-harvest processing.

Homegrown Company, Ltd., Kenya

Homegrown Company, Ltd., is a success story of production and export of packaged horticulture produce from Kenya. The company ventured into Kenya in 1982 and focused on the processing and export of vegetables to the UK market. The business strategy has been the production and packaging of produce at source so that it can be exported ready for the market outlet without further packaging abroad. To ensure the desired quality and supply of fresh produce, it was important for Homegrown to enter into partnerships with local farmers to complement its own production. Through this partnership the company is able to source about 25% of the total requirement from contracted farmers, and in some cases such as French beans, 100% of the total requirement.[15]

All farmers supplying to the Homegrown Company, Ltd., have a supply contract. The contract is explicit in terms of the commodity to be supplied, the period of supply, and the desired quality and quantity to be supplied. This implies that farmers on contract are able to work out their production schedules and put in place the necessary inputs to meet the contract quantities and quality. By implication also, farmers agree to follow the recommended crop husbandry so as to maintain the required quality. This contractual arrangement was initiated by the company as a strategy for achieving optimal resource use in the export of fresh produce from Kenya. Through this strategy, the company has its own nucleus of farm production units to meet a certain level of its requirements and a network of farmers contracted to provide the balance. Contracts entered with farmers for the supply of various types of fresh farm produce explicitly indicate the price as well as other quality dimensions that are important for delivery of the desired produce.

By entering into a supply contract, farmers enjoy the benefits of an assured market for their farm produce while at the same time benefiting from the fact that their farming activity risk is minimized by the certainty with which their production decisions are made. Farmers enjoy an assured price for the various grades of farm produce that they deliver to the contracting company. Due to the relative involvement of the contractor in the production process, farmers are supplied with the latest farming technology, such as the latest crop varieties and crop husbandry techniques. This has been particularly notable in the production of garden peas. The provision of technical extension by the contractor has played a key role in ensuring that farmers are able to optimize their production in terms of quality and quantity. Homegrown Company, Ltd., also supplies fertilizers, and agro-chemicals on credit to those farmers who need material credit so they can produce the expected quantities and qualities.

Sampa Jimini Cooperative Cashew Processing Society, Ghana

Sampa Jimini Cooperative Cashew Processing Society, located in Brong-Ahafo, was established in 1994 with the help of Technoserve, an American NGO. There are 18 workers, including one factory manager and two assistants. Membership of the Sampa Processing Society is 55. The society has elected executives and operates on the guidelines of a cooperative. In

1994, two processing societies were formed. The Department of Cooperatives provided the requisite training on the operation and management of the societies. In 1995, Technoserve sponsored training in processing in Nigeria and helped with the acquisition of equipment. Processing started in the year 2000.[16]

Four tons of raw nuts were processed into 1.14 tons of kernel. Between 2001 and 2002, 15 tons of raw nuts were purchased and processed. The kernels are sold to Golden Harvest Company, Ltd., in Accra. In 2002, the buyer started experiencing problems with the marketing of the kernels, which has affected prompt payment to the Society. As a result, the Society has looked for other marketing outlets such as the Indian community in Tamale and other sales outlets in Accra.

Vertical and horizontal farm-agribusiness linkages existed in cashew nut production. These linkages, however, were informal with no written contracts. Four types of linkages were functional at the time of the study: between farmers and the processing society; between society and Technoserve for business development services and technical advice; between the processing society and Golden Harvest Company, Ltd., for final processing; and between Golden Harvest and Technoserve for business development services, training, and technical advice. Farmers supply the processing society with raw nuts for processing. The processing society in turn educates the farmers on the best treatment and drying practices to get good nuts that attract a premium price. Technoserve encouraged the formation of the farmers' association and the processing society, organized training on cooperative organization, and introduced the society to financial institutions.

The linkage between the processing society and the marketing company has also been facilitated by Technoserve to ensure a ready market for the society's products. The connection is strengthened by the fact that the processing society owns 60 million shares of Golden Harvest, which in return provides training and information on international market developments. There were no contractual agreements. Technoserve played a significant role in establishing the connections observed in this case by introducing the concept of value addition by sponsoring training programs. Technoserve initiated all the linkages with the marketing firms and the other government institutions such as the Department of Cooperatives.

Blue Skies Agro-Processing Company, Ghana

Blue Skies Agro-processing Company, Ltd., is located about 25 km from Accra.[17] The company processes fresh fruits for supermarkets in some European markets. Fruits processed include pineapple, mangoes, watermelon, passion fruit, and pawpaw. While most fruit is procured in Ghana, supply gaps are filled by imports from South Africa, Egypt, Kenya, Brazil, and the UK. The company started with 38 workers and has since increased the workforce to 450, 60% of whom are permanent staff. The processed products of the company conform to the standards of the European Retailer Partnership Good Agricultural Practices (EUREGAP). In the last four years the company has grown tremendously, expanding its processing facilities. Through good extension services and training of farmers, coupled with higher price offers, the company rapidly increased its processing capacity from 1 ton per week to about 35 tons per week. Blue Skies is known to pay its farmers promptly and also to offer a higher price per kilogram of pineapple.

Farmers receive technical training and advice from the processing company free of charge to ensure that their produce meets the company's quality standards. Committed and loyal farmers can also purchase inputs and equipment interest free. Only farmers who are EUREPGAP certified are obliged to sell to the company because of the investment the company makes in getting farmers certified. There is a ready market for Blue

Skies' products in the EU market. The company is committed to supplying products on time and in the right quantities to supermarkets. To help its farmers, Blue Skies provides its dedicated farmers with credit and has worked to improve road infrastructure near farms and enhance access by company trucks.

Communication Technology in Food Processing

Conventional media, radio, and video are powerful, accessible, and relevant forces of agricultural innovation and transformation in Africa. Two-thirds of rural women creatively applied ideas illustrated by videos demonstrating improved food processing techniques compared to less than 20% who attended training workshops in Cotonou, Benin. The power of radio and video programming is not adequately recognized and accorded sufficient attention by Africa's policy makers, stifling the potential of these media to unleash farmer innovations. Farmers' innovations are often shaped by capital limitations and mainly rely on locally available resources, of which knowledge is key.

Video provides a powerful, low-cost medium for farmer-to-farmer extension and for exposing rural communities to new ideas and practices. A recent study examined the impacts of educational videos featuring early adopting farmers demonstrating the use of new technologies and techniques.[18] The study found that when women watched videos featuring fellow farmers demonstrating new techniques, they showed better learning and understanding of the technology and creatively applied its central ideas. Innovation levels of 72% were recorded in villages where videos were used to introduce women to improved rice processing techniques. This can be compared to 19% innovation among farmers who attended training workshops. When women who had attended training workshops watched the videos, the innovations increased to 92%. Watching videos spurred greater innovation than did conventional farmer training techniques. Notably high levels of creativity (67%) were recorded among women who did not have access to the rice processing technology featured in the video. The adaptations by Benin women to improve rice processing after having watched the video illustrate the power of video to quickly stimulate creativity among rural people, who are often seen as much more passive technology consumers. Besides being more powerful, video may also be able to reach more people than conventional training workshops. Drawing lessons from a similar rural learning initiative undertaken in Bangladesh, the Africa Rice Center with a wide range of partners is using local language videos to train farmers on various facets of rice production and processing in Benin, Ethiopia, Gambia, Ghana, Nigeria, and Senegal, among other countries.

By 2009, the rice videos had been translated into 30 African languages and were being used by more than 400 community-based organizations across Africa to strengthen their own capacity in rice technologies.[19] The videos, which are disseminated through mobile cinema vans or local organizations, have been viewed by about 130,000 farmers across Africa, reaching three times as many farmers as face-to-face farmer training workshops. Partner organizations in various countries are combining the videos with radio programming to reinforce the lessons and knowledge.

In Guinea, one radio station, Radio Guinée Maritime, has aired interviews with farmers involved in this program reaching some 800,000 listeners, an experience that has been replicated in Gambia, Nigeria, and Uganda. To effectively capitalize on the potential of radio and video technologies in Africa, proponents advise broadening the dissemination of innovations beyond those developed by the traditional research and extension systems to include localized farmer innovations also.

Social Entrepreneurship and Local Innovations

Social enterprises are emerging as major economic players worldwide.[20] Their role in African agriculture is growing. An example of such an initiative is the One Acre Fund, a non-profit organization based in Bungoma (western Kenya) that provides farmers with the tools they need to improve their harvests and feed their families.[21] Life-changing agricultural technologies already exist in the world; One Acre Fund's primary focus is on how to distribute these technologies in a "farmer-usable" way, and how to get farmers to permanently adopt these technologies. One Acre Fund currently serves 23,000 farm families (with 115,000 children in those families) in Kenya and Rwanda.

From the beginning, One Acre Fund talked to farmers to understand what they need to succeed: finance, farm inputs, education, and access to markets. One Acre Fund offers a service model that addresses each of these needs. When a farmer enrolls with One Acre Fund, she joins as part of a group of 6 to 12 farmers. She receives an in-kind loan of seed and fertilizer, which is guaranteed by her group members. One Acre Fund delivers this seed and fertilizer to a market point within two kilometers of where she lives and a field officer provides in-field training on land preparation, planting, fertilizer application, and weeding. These trainings are stan-dardized across One Acre Fund's entire operation and include interactive exercises, simple instructions, and group modeling of agriculture techniques. For instance, after a field officer teaches a group of farmers how to use a planting string to space rows of crops, he asks them to model the technique in the field so that he can offer immediate feedback.

Over the course of the season, the field officer monitors the farmer's fields. At the end of the season, he trains her on how to harvest and store her crop. One Acre Fund also offers a harvest buy-back program that farmers can participate in if they choose. Final loan repayment is several weeks after harvest—98% of farmers repay their loans.

Before joining One Acre Fund, many farmers in Kenya were harvesting five bags of maize from half an acre of land. After joining One Acre Fund, their harvests typically increase to 12 to 15 bags of maize from the same half acre of land. This represents a doubling in farm profit per planted acre—twice as much income from the same amount of land.

The field officer is the most important part of the One Acre Fund service model. These field officers are typically recruited from the communities in which they work. One Acre Fund consciously chooses not to hire university-educated horticulturists for its field staff (like most NGOs) because most of the information farmers need to know is encapsulated in a few simple lessons. The Fund prefers to employ down-to-earth, hardworking staff—many of whom are farmers themselves—who have strong leadership potential within their own communities.

One Acre Fund's field officers each work with roughly 120 farmers, and they visit each of their farmers on a weekly or biweekly basis. At these meetings, they conduct trainings, check germination rates, troubleshoot problems in the field, and collect repayment. Over the course of the season, One Acre Fund's field officers cultivate a strong bond with their farmers. They respect the feedback their farmers give them about farming techniques and One Acre Fund's program. In turn, One Acre Fund farmers have a deep appreciation for how knowledgeable their field officers are and how hard they work to serve their customers. Many farmers call their field officers "teacher."

Local innovations represent one of Africa's least recognized assets. For more than 20 years India's Honey Bee Network and

Society for Research and Initiatives for Sustainable Technologies and Institutions have been scouting for innovations developed by artisans, children, farmers, women, and other community actors. They have built a database of more than 10,000 innovations.[22]

To further the work, India's Department of Science and Technology created the National Innovation Foundation (NIF) in 2000. Its aim is "providing institutional support in scouting, spawning, sustaining and scaling up grassroots green innovations and helping their transition to self supporting activities." NIF has so far filed over 250 patent applications for the ideas in India, of which 35 have been granted. Another seven applications have been filed in the United States of which four have been granted.

To facilitate the commercialization and wider application of the innovations, NIF works with institutions such as the Grassroots Innovations Augmentation Network, which serves as a business incubator. Some of the objectives of the Honey Bee Network are now part of the work of the Prime Minister's National Innovation Council. Africa's diversity in agricultural and ecological practices offers unique opportunities for creative responses to local challenges. Such responses form a foundation upon which to supplement formal institutions with entrepreneurial activities driven by local innovations.

Conclusions

Despite strong growth in the private seed sector in Africa over the last decade, most of Africa's millions of small-scale farmers lack easy access to affordable, high-quality seeds. Seed policies and regulations currently differ across African countries, limiting opportunities for trade and collaboration. However, efforts are under way to develop regional trading blocs in the seed industry. For example, across the 14 Southern Africa Development Community (SADC) countries, seed industry stakeholders have been formulating a single policy document to enable companies to move seed and breeding material across national borders, register varieties more easily, and market their products regionally. A parallel initiative is under way for East African Community (EAC) countries. These efforts need to be finalized in east and southern Africa, replicated across all Africa's subregional organizations, and complemented by parallel efforts in the African Union.

Like the formation of African seed companies, the creation and spread of value-added food processing enterprises could help African farmers retain a higher portion of the profits from the materials they produce. Food processing could also help reduce the threat of hunger by increasing the number of protein- and vitamin-rich products provided by the local market, as well as improve local incomes by tapping into international markets to get much needed export revenues from agriculture. Unlike the situation with seeds, growth in food processing will require fewer changes in government and regional policies. The key change will need to come in the areas of capital, so it is easier for individuals and companies to invest in the infrastructure, equipment, and training necessary to enter the food processing industry.

7

GOVERNING INNOVATION

African countries are increasingly focusing on promoting regional economic integration as a way to stimulate economic growth and expand local markets. Considerable progress has been made in expanding regional trade through regional bodies such as the Common Market for Eastern and Southern Africa (COMESA) and the East African Community (EAC). There are six other such Regional Economic Communities (RECs) that are recognized by the African Union as building blocks for pan-African economic integration. So far, regional cooperation in agriculture is in its infancy and major challenges lie ahead. This chapter will explore the prospects of using regional bodies as agents of agricultural innovation through measures such as regional specialization. The chapter will examine ways to strengthen the role of the RECs in promoting innovation. It adopts the view that effective regional integration is a learning process that involves continuous institutional adaptation.[1]

Through extensive examples of initiatives at the national or cross-border levels, this chapter provides cases for regional collaboration or scaling up national programs to regional programs. Africa's RECs have convening powers that position them as valuable vehicles. That is, they convene meetings of political leaders at the highest level, and these leaders take decisions that are binding on the member states; the member states then regularly report on their performance regarding these decisions. Such meetings provide good platforms for sharing information and best practices. Africa's RECs have established and continue to designate centers of excellence in various areas. COMESA, for instance, has established reference laboratories for animal and plant research in Kenya and Zambia. Designation of centers of excellence for specific aspects of agricultural research will greatly assist specialization within the RECs and put to common use the knowledge from the expertise identified in the region.

Regional Innovation Communities

Regional integration is a key component of enabling agricultural innovation because it dismantles three barriers to development: "weak national economies; a dependence on importing high-value or finished goods; and a reliance on a small range of low-value primary exports, mainly agriculture and natural resources."[2]

Physical infrastructure creates a challenge for many African countries but also presents an opportunity for the RECs to collaborate on mutually beneficial projects. In many parts of Africa, poor road conditions prevent farmers from getting to markets where they could sell their excess crops profitably. Poor road conditions include the lack of paved roads, the difficulty of finding transportation into market centers, and the high cost of having to pay unofficial road fees to either customs officials or other agents on the roads. These difficulties become more extreme when farmers have to get their crops across international borders to reach markets where sales are profitable.

The inability to sell crops, or being forced to sell them at a loss because of high transportation costs, prevents farmers from making investments that would increase the quantity and quality of their production, since any increase will not add to their own well-being, and the excess crops may go to waste. This is a problem where national governments and regional cooperation offer the best solution. Regional bodies, with representation from all of the concerned countries, are placed to address the needs for better subregional infrastructure and standardization of customs fees at only a few locations.

Having countries come together to address problems of regional trade, and particularly including representatives from both the private and public sectors, allows nations to identify and address the barriers to trade. The governments are now working together to address the transportation problem and to standardize a regional system of transport and import taxes that will reduce the cost of transporting goods between nations. This new cooperation will allow the entire region to increase its food security by capitalizing on the different growing seasons in different countries and making products available in all areas for longer periods of time, not just the domestic season. Such cooperation also provides African farmers with access to international markets that they did not have before, since it allows them to send their goods to international ports, where they can then sell them to other nations.

Regional economic bodies provide a crucial mechanism for standardizing transport procedures and giving farmers a chance to earn money selling their products. This cooperation works best when it happens both between countries and between the private and public sectors. Having multiple actors involved allows for better information, more comprehensive policy making, and the inclusion of many stakeholders in the decision-making process. Governments and private actors should strengthen their participation in regional bodies and use those groups to address transportation issues, market integration, and infrastructure problems.

Facilitating regional cooperation is emerging as a basis for diversifying economic activities in general and leveraging international partnerships in particular.[3] Many of Africa's individual states are no longer viable economic entities; their future lies in creating trading partnerships with neighboring countries. Indeed, African countries are starting to take economic integration seriously.[4] For example, the re-creation of the EAC is serving not only as a mechanism for creating larger markets but also is promoting peace in the region. Economic asymmetry among countries often is seen as a source of conflict.[5] However, the inherent diversity serves as an incentive for cooperation.

Emerging Regional Research Cooperation Trends

The original Organisation of African Unity (OAU) was transformed and restructured to form the African Union (AU) with a stronger mandate to ensure the socioeconomic development of the continent. The secretariats of the AU Commission and the New Partnership for Africa's Development (NEPAD) were divided into departments that reflect development priorities of the AU. For example, both have offices of agriculture and science and technology (S&T). The AU Commission takes responsibility for formulating common policies and programs of the AU and presents them for approval by heads of state and government. NEPAD, on the other hand, spearheads the implementation of approved policies and programs.

In 2006, heads of state and government approved the Consolidated Plan for African Science and Technology, which is both a policy and a program document for the promotion of S&T in Africa. Since then the AU Commission's Department of Human Resources, Science, and Technology has mobilized participation of member states in a specialized ministerial structure that advises AU heads of states and government on matters related to S&T. The NEPAD office of S&T managed the implementation of the research program that consisted of flagship programs covering themes such as biosciences, biodiversity, biotechnology, energy, water, material sciences, information and communication technology, space science, and mathematics.

Of the flagship programs, the biosciences program is the most advanced in that it has five regional programs that were developed bottom-up by scientists in the regions. The program has attracted loyal sponsors and some programs are at a stage of transforming their results into promising products. For example, Southern Africa Network for Biosciences (SANBio), which had identified HIV/AIDS as one of its priorities, has been conducting research into the validation of traditional medicines for affordable treatment of HIV/AIDS and HIV-related opportunistic infections. The study obtained very encouraging results. Two peptides with anti-HIV activity have been isolated and currently the study is focusing on developing this lead into a herbal medicine that could benefit HIV/AIDS patients.

Other regions have equally advanced programs that are derived from regional interests. However valuable as these programs appear, they have not captured the interest of governments and local industry enough to leverage local investment. For example, the SANBio program runs on the basis of Finnish and South African funding. This is a major limitation to growth of the programs. An analysis of the problem seems to reveal that the NEPAD science and technology program is bedeviled by its isolation from the local economy—the biosciences-related industry that has both government and related industry vested interests backed by local investment. At this stage, based on the absence of investment in the program, the NEPAD science and technology effort seems divorced from the African reality even though it is framed around African S&T-related problems.

Various research and development programs are being carried out in each of the networks created under NEPAD, such as Biosciences Eastern and Central Africa Network (BecANet), West African Biosciences Network (WABNet), and North Africa Biosciences Network (NABNet). Research programs at SANBio include technology transfer to local communities, especially women, for producing mushrooms using affordable local resources and research in aquaculture systems involving use of plastic sheeting as covers to improve productivity of pond environments for increased fish growth.

BecANet has implemented three flagship and five competitive grant projects in the areas of banana, sorghum, bovine and human tuberculosis, teff, cassava, sweet potatoes, and tsetse and trypanosomiasis. The BecANet hub hosted at least 24 research projects on crops and livestock. The research on trypanosomiasis led to the discovery of a compound that could be used to produce a drug for treating sleeping sickness. This compound was subsequently patented. The results from the work on tracking bush meat in Kenya using DNA techniques at the BecANet hub suggest that bush meat, as whole meat, is not sold within Nairobi, but outside the city. This type of study has wider implications for food recall and traceability in Africa.

WABNet has implemented a flagship project on the inventory and characterization of West African sorghum genetic resources. This project conducted a germplasm collection expedition during which 45 new sorghum accessions were collected from 15 districts including (Upper West Region), Bawku (Upper East Region), and Tamale (Northern Region) of Ghana by November 2009. In Ghana, for example, 245 accessions of sorghum were at a risk of being lost as a result of deteriorating storage conditions in the gene banks. This project was instrumental in recovering about 100 of these accessions. This project had to therefore refocus its objectives on the new collections and on those that had been recovered from the banks. Sorghum nutritional enhancement through mutagenesis and genomics is inducing mutagenesis through irradiation, with the aim of producing new varieties of sorghum that could be higher yielding and more nutritionally enhanced. Trials on this aspect of the project are still ongoing.

NABNet carried out a flagship project on production of elite biofortified North African barley genotypes tolerant to biotic and abiotic stresses. In addition, the network is implementing the following research projects: multidisciplinary investigation of the genetic risk factors of type II diabetes and its complications in North Africa; biotechnological approaches to protect date palm against major plagues in North Africa; and production of *Bt* bioinsecticides useful for biological control of the phytopathogenic insects and human disease vectors.

Within SANBio, through support from the governments of Finland and South Africa, NEPAD African Biosciences Initiative (NEPAD/ABI) is establishing a bioinformatics center at the University of Mauritius and an indigenous knowledge systems center at the University of North West in South Africa. These centers will enhance capacity building in these domains in southern Africa.

Through the support of the government of Canada and International Livestock Research Institute (ILRI), laboratories at ILRI have been upgraded offering opportunities to African scientists to carry out cutting-edge biosciences research at the BecANet hub. As the facilities became fully operational by the end of 2009, it is expected that there will be an increase in the number of projects being carried out in 2010 onward. The infrastructure upgraded at the BecANet hub can support research in bioinformatics, diagnostics, sequencing, genotyping, molecular breeding, and genetic engineering of crops, livestock, and wild animals and plants.

Within WABNet, NEPAD/ABI has established and furnished offices in Dakar and Senegal and a biotechnology laboratory at the University of Ouagadougou in Burkina Faso. These facilities will enhance coordination, research, and development in the region.

Networking is viewed as an essential component for accelerating capacity building and collaborative research efforts among institutions involved in biosciences in Africa. More endowed biosciences institutions are being linked to less endowed ones in all the regions of the continent. Over the years, thematic research teams have been established, such as ones for improving banana production; for validation of herbal remedies, bioinformatics, fisheries, and aquaculture; and for cereal improvement, just to mention a few. A more important development is that thematic networks are now running across regional and international boundaries and broadening the scope for participating international research programs.[6]

For example, the trypanosomiasis and tsetse fly research network consists of institutions and researchers in the SANBio and BecANet countries. Networking has provided opportunities for young men and women to access training facilities in several countries and subregions. A total of 20 students from post-conflict countries of the Democratic Republic of the Congo, Sudan, Congo Brazzaville, and Burundi are now training for postgraduate qualifications in Kenya and Uganda.

The COMESA region is home to some of Africa's most important fisheries resources. These include, among others, marine systems in the western Indian Ocean, the southeastern Atlantic, the Mediterranean, and the vast freshwater systems of the Nile, Congo, and Zambezi Basins and the Great Lakes found within them. Taken together, COMESA member states have access to a coastline of some 14,418 kilometers, a continental shelf of about 558,550 square kilometers, and a total Exclusive Economic Zone area of about 3.01 million square kilometers, as well as inland waters of about 394,274 square kilometers.

COMESA has identified aquaculture as a priority growth area for achieving the objectives and targets of the Comprehensive Africa Agriculture Development Programme (CAADP) to significantly contribute to food and nutrition security in the region. In September 2008, member states agreed on the outline of a COMESA Fisheries and Aquaculture Strategy to enable COMESA to scale up their coordinating and facilitative role.

COMESA in partnership with Lake Victoria Fisheries Organization is implementing a program that seeks to maximize economic benefits and financial returns on processed Nile perch through innovation and development of sophisticated value-added products. Currently, with decreasing levels of Nile perch, fish processing establishments are operating well below capacity, sometimes as low as 20% capacity. To remain competitive, companies are seeking to transform by-products or fish fillet waste into high-value products for export through use of innovative processes and production methods.

Three companies are supported by COMESA to achieve this. Interventions have focused on two areas. First, they support companies to develop new products through piloting of new recipes, processes, and production methods while utilizing new or improved technologies. Second, they help fish processing companies to assess food safety aspects of new products developed from fish wastes and the application of appropriate quality assurance and food safety systems based on the Hazard Analysis and Critical Control Point (HACCP) system to address the associated risks. The HACCP is a food and pharmaceutical safety approach that addresses physical, chemical, and biological hazards as a preventive measure rather than inspection of finished products.[7]

The results are encouraging. One company successfully launched frozen fish burgers and established HACCP-based food safety systems that are compliant with EU requirements. As a result, the company has accessed high-value markets in the European Community (EC).

Another product development innovation in the fisheries sector is using Nile perch skin as a raw material for exotic leather products. This has been picked up by COMESA, which has developed a value chain sector strategy for leather and leather products, part of which is now being implemented under a Canadian-funded Programme for Building African Capacity for Trade. A key issue for the region's competitiveness is innovation in new products and markets for leather products rather than continuing to export raw hides and skins. The fish skin niche market that was identified started in Uganda, at the Crane Shoes factory in Kampala. The use of fish skin began as the factory sought new markets and alternatives to leather to compete with Chinese imports. The key question, however, is how to support the industry to achieve long-term sustainability, which may require a rare form of aquaculture for the Nile perch, which is becoming a rare species in Lake Victoria.

An investment study on Uganda's fish and fish farming industry identified a number of policy measures that are being put in place in order to benefit from the fisheries sector. These include a tripartite Lake Victoria Environment Management Project established to ensure improved productivity of Lake Victoria; support for the Aquaculture Research and Development Institute at Kajjansi; provision of resources to upgrade landing sites and quality control laboratories to meet international standards; and provision of resources to strengthen the Uganda National Bureau of Standards and Inspectorate section of the Fisheries Department.

Fostering the Culture of Innovation

Improving the Governance of Innovation

Promoting a growth-oriented agenda will require adjustments in the structure and functions of government at the regional, national, and local levels. Issues related to science, technology, and innovation must be addressed in an integrated way at the highest possible levels in government. There is therefore a need to strengthen the capacity of presidential offices to integrate science, technology, and innovation in all aspects of government. In 2010 no African head of state or government had a chief scientific adviser.

The intensity and scope of coordination needed to advance agricultural innovation exceeds the mandate of any one ministry or department. As noted elsewhere, Malawi addressed the challenge of coordination failure by presidential control of agricultural responsibilities. The need for high-level or executive coordination of agricultural functions is evident when one takes into account the diverse entities that have direct relevance to any viable programs. Roads are important for agriculture, yet they fall under different ministries that may be more concerned with connecting cities than rural areas. Similarly, ministries responsible for business development may be focusing on urban areas where there is a perception of short-term returns to investment. The point here is not to enter the debate on the so-called urban bias.[8]

The main point is to highlight the importance of strategic coordination and alignment of the functions of government to reflect contemporary economic needs. Aligning the various organs of government to focus on the strategic areas of economic efforts requires the use of political capital. In nearly all systems of government such political capital is vested in the chief executive of a country, either the president or the prime minister depending on the prevailing constitutional order. It would follow from this reasoning that presidents or prime ministers should have a critical agricultural coordination role to perform. They can do so by assuming the position of minister or by heading a body charged with agricultural innovation. The same logic also applies for the RECs.

The dominant thinking is to create "science and innovation desks" in the RECs. Such desks will mirror the functions of the science and technology ministries at the national level. It is notable that currently no African leaders are supported by effective mechanisms that provide high-level science, technology, and engineering advice. The absence of such offices (with proper terms of reference, procedures, legislative mandates, and financial resources) hampers the leaders' ability to keep abreast of emerging technological trends and to make effective decisions. Rapid scientific advancement and constant changes in the global knowledge ecology require African leaders at all levels (heads of RECs, presidents, or prime ministers and heads of key local authorities such as states or cities) to start creating institutions for science advice. In 2010 COMESA led the way by adopting decisions on science, technology, and innovation along these lines (see Appendix II). Agricultural innovation could be the first beneficiary of informed advice from such bodies.

Bringing science, technology, and engineering to the center of Africa's economic renewal will require more than just political commitment; it will take executive leadership. This challenge requires concept champions, who in this case will be heads of state spearheading the task of shaping their economic policies around science, technology, and innovation. So far, most African countries have failed to develop national policies that demonstrate a sense of focus to help channel emerging technologies into solving developmental problems. They still rely on generic strategies dealing with "poverty alleviation" without serious consideration of the sources of economic growth.

One of the central features of executive guidance is the degree to which political leaders are informed about the role of science, technology, and engineering in development. Advice on science, technology, and innovation must be included routinely in policy making. An appropriate institutional framework must be created for this to happen. Many African cabinet structures are merely a continuation of the colonial model, structured to facilitate the control of local populations rather than to promote economic transformation.

Advisory structures differ across countries. In many countries, science advisers report to the president or prime minister, and national scientific and engineering academies provide political leaders with advice. Whatever structure is adopted, the advising function should have some statutory mandate to advise the highest levels of government. It should have its own operating budget and a budget for funding policy research. The adviser should have access to good and credible scientific or technical information from the government, national academies, and international networks. The advisory processes should be accountable to the public and be able to gauge public opinion about science, technology, and innovation.

Successful implementation of science, technology, and innovation policy requires civil servants with the capacity for policy analysis—capacity that most current civil servants lack. Providing civil servants with training in technology management, science policy, and foresight techniques can help integrate science, technology, and innovation advice into decision making. Training diplomats and negotiators in science, technology, and engineering also can increase their ability to discuss technological issues in international forums.

African countries have many opportunities to identify and implement strategic missions or programs that promote growth through investments in infrastructure, technical training, business incubation, and international trade. For example, regional administrators and mayors of cities can work with government, academia, industry, and civil society to design missions aimed at improving the lives of their residents. Universities located in such regions and cities could play key roles as centers of expertise, incubators of businesses, and overall sources of operational outreach to support private and public sector activities. They could play key roles in transferring technology to private firms.[9]

Similar missions could be established in rural areas. These missions would become the organizing framework for fostering institutional interactions that involve technological learning and promote economies of scale. In this context, missions that involve regional integration and interaction should be given priority, especially where they build on local competencies.

This approach can help the international community isolate some critical elements that are necessary when dealing with such a diverse set of problems as conservation of forests, provision of clean drinking water, and improving the conditions of slum dwellers. In all these cases, the first major step is the integration of environmental considerations into development activities.

Reforming the Structures of Innovation Governance

The RECs offer a unique opportunity for Africa to start rethinking the governance of innovation so that the region can propel itself to new frontiers and run its development programs in an enlightened manner that reflects contemporary challenges and opportunities. The focus of improvements in governance structures should be at least in four initial areas: a high-level committee on science, innovation, technology, and engineering; regional science, technology, and engineering academies; an office of science, technology, and innovation; and a graduate school of innovation and regional integration.

Committee on science, innovation, technology, and engineering: The committee will be a high-level organ of each REC that will report directly to the councils of ministers and presidential summits. Its main functions should be to advise the respective REC on all matters pertaining to science, technology, engineering, and innovation. The functions should include, but not be limited to, regional policies that affect science, technology, engineering, and innovation. It shall also provide scientific and technical information needed to inform and support public policy on regional matters in areas of the competence of the RECs (including economy, infrastructure, health, education, environment, security, and other topics).

For such a body to be effective, it will need to draw its membership from a diversity of sectors including government, industry, academia, and civil society. The members shall serve for a fixed term specified at the time of appointment. Within these sectors, representation should reflect the fact that science, technology, engineering, and innovation are not limited to a few ministries or departments but cover the full scope of the proper functioning of society.

The committee should meet as needed to respond to information requests by chief executive, councils of ministers, or the summits. To meet this challenge, the committee should solicit information from a broad spectrum of stakeholders in the research community, private sector, academia, national research institutes, government departments, local government, development partners, and civil society organizations. The committee's work can be facilitated through working groups or task forces set up to address specific issues.

The committee's work will be supported by the Office of Science, Innovation, Technology, and Engineering headed by a director who also serves as the Chief Science, Innovation, Technology and Engineering Adviser to the chief executive. A national analogue of such a committee is India's National Innovation Council that was established in 2010 by the Prime Minister to prepare a roadmap for the country's Decade of Innovation (2010–2020). The aim of the council is to develop an Indian innovation model that focuses on inclusive growth and the creation of institutional networks that can foster inclusive innovation. The council will promote the creation of similar bodies at the sectoral and state levels.[10]

Regional academies of science, innovation, technology, and engineering: African countries have in recent years been focusing on creating or strengthening their national academies of science and technology. So far 16 African countries (Cameroon, Egypt, Ethiopia, Ghana, Kenya, Madagascar, Mauritius, Morocco, Mozambique, Nigeria, Senegal, South Africa, Sudan, Tanzania,

Uganda, and Zimbabwe) have national academies. There is also the nongovernmental African Academy of Sciences (AAS). It is notable that despite Africa's growing emphasis on investing in infrastructure, only one African country (South Africa) has an academy devoted to promoting engineering. Most of the others seek to recognize engineers through regular scientific academies, but their criteria for selection tend to focus on publications and not practical achievements. A case can be made on the need to expand the role of academies in providing advice on engineering-related investments.

The creation of regional academies of science, innovation, technology, and engineering will go a long way in fostering the integration of the various fields and disciplines so that they can help to foster regional integration and development.[11] The main objectives of such academies would be to bring together leaders of the various regions in science, innovation, technology, and engineering to promote excellence in those fields. Their priorities would be to strengthen capabilities, inspire future generations, inform public debates, and contribute to policy advice.

The fellows of the academies will be elected through a rigorous process following international standards adopted by other academies. Their work and outputs should also follow the same standards used by other academies. The academies should operate on the basis of clear procedures and should operate independently. They may from time to time be asked to conduct studies by the RECs but they should also initiate their own activities, especially in areas such as monitoring scientific, technological, and engineering trends worldwide and keeping the RECs informed about their implications for regional integration and development. Unlike the committee, the academies will operate independently and their advisory functions are only a part of a larger agenda of advancing excellence in science, innovation, technology, and engineering.

As part of their public education mission, the academies could collaborate with groups such as the Forum for Agricultural Research in Africa, an umbrella organization that brings together and forges coalitions of leading stakeholders in agricultural research and development in Africa. If needed, specialized regional academies of agriculture could be created to serve the sector. Such agricultural academies could benefit from partnerships with similar organizations in countries such as China, India, Sweden, and Vietnam. The proposed academies will need to work closely with existing national academies and AAS.

Office of science, innovation, technology, and engineering: The RECs will need to create strong offices within their secretariats to address issues related to science, innovation, technology, and engineering. The bulk of the work of such offices will be too coordinate advisory input as well as serve as a link between the various organs of the RECs and the rest of the world. The head of the office will have two main functions. First, the person will serve as the chief adviser to the various organs of the RECs (through the chief executive). In effect, the person will be the assistant to the chief executive on science, innovation, technology, and engineering. Second, the person will serve as director of the office and be its representative when dealing with other organizations. In this role the director will be a promoter of science, innovation, technology, and engineering whereas in the first role the person will serve as an internal adviser.

For such an office to be effective, it will need to be adequately funded and staffed. It can draw from the personnel of other departments, academies, or organizations to perform certain duties. In addition to having adequate resources, the office will need to develop transparent procedures on how it functions and how it relates to other bodies. It is imperative that the functions of the office be restricted to the domain of advice and it should not have operational responsibilities, which belong to the national level.

School of regional integration: The need to integrate science and innovation in regional development will require the creation of human capacity needed to manage regional affairs. So far, the RECs rely heavily on personnel originally trained to manage national affairs. There are very few opportunities for training people in regional integration. Ideally, there should be a graduate African School of Regional Integration to undertake research, professional training, and outreach on how to facilitate regional integration. Such a school could function either as a stand-alone institution or in conjunction with existing universities. A combination of the two where an independent school serves as a node in a network of graduate education in regional integration is also an option.

The school could focus on providing training on emerging issues such as science and innovation. It can do so through short executive courses, graduate diplomas, and degree programs. There is considerable scope for fostering cooperation between such a school and well-established schools of government and public policy around the world. The theme of regional integration is a nascent field with considerable prospects for growth. For this reason it would not be difficult to promote international partnerships that bring together regional and international expertise.

The school could also serve as depository of knowledge gained in the implementation of regional programs. Staff from the RECs could serve as adjunct faculty and so could join it as full-time professors of the practice of regional integration. The school could also work with universities in the region to transfer knowledge, curricula, and teaching methods to the next generation of development practitioners. The area of agricultural innovation would be ideal for the work of such a school and a network of universities that are part of the regional innovation system.

Funding Innovation

One of the key aspects of technological development is funding. Financing technological innovation should be considered in the wider context of development financing. Lack of political will is often cited as a reason for the low level of financial support for science, technology, and innovation in Africa. But a large part of the problem can be attributed to tax and revenue issues that fall outside the scope of science and technology ministries.[12] For example, instruments such tax credits that have been shown to increase intensity of research and development activities are unlikely to work in policy environments without a well-functioning tax regime.[13] Other instruments such as public procurement can play a key role in stimulating innovation, especially among small and medium-sized enterprises (SMEs).[14]

Currently, Africa does not have adequate and effective mechanisms for providing support to research. Many countries have used a variety of models, including independent funds such as the National Science Foundation in the United States and the National Research Fund of South Africa. Others have focused on ensuring that development needs guide research funding and, as such, have created specific funding mechanisms under development planning ministries. While this approach is not a substitute for funding to other activities, it distinguishes between measures designed to link technology to the economy from those aimed at creating new knowledge for general learning. What is critical, however, is to design appropriate institutional arrangements and to support funding mechanisms that bring knowledge to bear on development.

Creating incentives for domestic mobilization of financial resources as a basis for leveraging external support would be essential. Other innovations in taxation, already widespread around the world, involve industry-wide levies to fund research, similar to the Malaysian tax mechanism to fund research. Malaysia imposed cesses on rubber, palm oil, and timber to fund the Rubber Research Institute, the Palm Oil Research Institute, and the Forestry Research Institute. A tax on tea helps fund research on and marketing of tea in Sri Lanka. Kenya levies a tax on its tea, coffee, and sugar industries, for example, to support the Tea Research Foundation, the Coffee Research Foundation, and the Kenya Sugar Board. These initiatives could be restructured to create a funding pool to cover common areas. Reforming tax laws is an essential element in the proposed strategy. Private individuals and corporations need targeted tax incentives to contribute to research funds and other technology-related charitable activities. This instrument for supporting public welfare activities is now widely used in developing countries. It arises partly because of the lack of experience in managing charitable organizations and partly because of the reluctance of finance ministries to grant tax exemptions, fearing erosion of their revenue base.

The enactment of a foundation law that provides tax and other incentives to contributions to public interest activities, such as research, education, health, and cultural development, would promote social welfare in general and economic growth in particular. Other countries are looking into using national lotteries as a source of funding for technological development. Taxes on imports could also be levied to finance innovation activities, although the World Trade Organization may object to them. Another possibility is to impose a tax of 0.05% or 0.1% of the turnover of African capital markets to establish a global research and development fund, as an incentive for them to contribute to sustainable development.

Other initiatives could simply involve restructuring and redefining public expenditure. By integrating research and development activities into infrastructure development, for example, African governments could relax the public expenditure constraints imposed by sectoral budgetary caps. Such a strategy has the potential to unlock substantial funds for research and development in priority areas. But this strategy requires a shift in the budgetary philosophy of the international financial institutions to recognize public expenditures on research and development as key to building capabilities for economic growth.

Financing is probably one of the most contentious issues in the history of higher education. The perceived high cost of running institutions of higher learning has contributed to the dominant focus on primary education in African countries. But this policy has prevented leaders from exploring avenues for supporting higher technical education.

Indeed, African countries such as Uganda and Nigeria have considered new funding measures including directed government scholarships and lowering tuition for students going into the sciences. Other long-term measures include providing tax incentives to private individuals and firms that create and run technical institutes on the basis of agreed government policy. Africa has barely begun to utilize this method as a way to extend higher technical education to a wider section of society. Mining companies, for example, could support training in the geosciences. Similarly, agricultural enterprises could help create capacity in business.

Institutions created by private enterprises can also benefit from resident expertise. Governments, on the other hand, will need to formulate policies that allow private sector staff to serve as faculty and instructors in these institutions. Such programs also would provide opportunities for students to interact with practitioners in addition to the regular faculty.

Much of the socially responsible investment made by private enterprises in Africa could be better used to strengthen the continent's technical skill base. Additional sources of support could include the conversion of the philanthropic arms of various private enterprises into technical colleges located in Africa.

Governmental and other support will be needed to rehabilitate and develop university infrastructures, especially information and communications facilities, to help them join the global knowledge community and network with others around the world. Such links will also help universities tap into their experts outside the country. Higher technical education should also be expanded by creating universities under line ministries as pioneered by telecom universities such as the Nile University (Egypt), the Kenya Multimedia University, and the Ghana Telecoms University College. Other line ministry institutions such as the Digital Bridge Institute in Nigeria are also considering becoming experiential universities with strong links with the private sector.

Governments and philanthropic donors could drive innovation through a new kind of technology contest.[15] One approach is to offer proportional "prize rewards" that would modify the traditional winner-take-all approach by dividing available funds among multiple winners in proportion to measured achievement.[16] This approach would provide a royalty-like payment for incremental success.[17]

Promoting innovation for African farmers has proven especially challenging, due to a wide variety of technological and institutional obstacles. A proportional-prize approach is particularly suited to help meet the needs of African farmers. For that purpose a specific way should be devised to implement prize rewards, to recognize and reward value creation from new technologies after their adoption by African farmers.

In summary, the effectiveness of innovation funding depends on choosing the right instrument for each situation—and perhaps, in some situations, developing a new instrument that is specifically suited to the task. Prizes are distinctive in that they are additional and temporary sources of funding, they are used when needed to elicit additional effort, and they can reveal the most successful approaches for reaching a particular goal. For this reason, a relatively small amount of funding in a well-designed prize program can help guide a much larger flow of other funds, complementing rather than replacing other institutional arrangements.

Available evidence suggests that investments in agricultural research require long-term sustained commitment. This is mainly because of the long time lags associated with such investments, ranging from 15 to 30 years taking into account the early phases of research.[18] Part of the time lag, especially in areas such as biotechnology, is accounted for by delays in regulatory approvals or the high cost of regulation. This is true even in cases where products have already been approved and are in use in technology pioneering countries.[19] These long time lags are also an expression of the fact that the economic systems co-evolve with technology and the process involves adjustments in existing institutions.[20]

Joining the Global Knowledge Ecology

Leveraging Africa's Diasporas

Much of the technological foundation needed to stimulate African development is based on ideas in the public domain (where property rights have expired). The challenge lies in finding ways to forge viable technology alliances.[21] In this regard, intellectual property offices are viewed as important sources of information needed for laying the basis for technological innovation.[22] While intellectual property protection is perceived as a barrier to innovation, the challenges facing Africa lie more in the need to build the requisite human and institutional capability to use existing technologies. Much of this can be achieved though collaboration with leading research firms and product development.[23] This argument may not hold in regard to emerging fields such as genomics and nanotechnology.

One of the concerns raised about investing in technical training in African countries is the migration of skilled manpower to industrialized countries.[24] The World Bank has estimated that although skilled workers account for just 4% of the sub-Saharan labor force, they represent some 40% of its migrants.[25] Such studies tend to focus on policies that seek to curb the so-called brain drain.[26] But they miss the point. The real policy challenge for African countries is figuring out how to tap the expertise of those who migrate and upgrade their skills while out of the country, not engage in futile efforts to stall international migration.[27] The most notable case is the Taiwanese diaspora, which played a crucial role in developing the country's electronics industry.[28] This was a genuine partnership involving the mobility of skills and capital.

Countries such as India have understudied this model and come to the conclusion that one way to harness the expertise is to create a new generation of "universities for innovation" that will seek to foster the translation of research into commercial products. In 2010 India unveiled a draft law that will provide for the establishment of such universities. The law grew out India's National Knowledge Commission, a high-level advisory body to the Prime Minister aimed at transforming the country into a knowledge economy.[29]

A number of countries have adopted policy measures aimed at attracting expatriates to participate in the economies of their countries of origin. They are relying on the forces of globalization such as connectivity, mobility, and interdependence to promote the use of the diaspora as a source of input into national technological and business programs. These measures include investment conferences, the creation of rosters of experts, and direct appeals by national leaders. It is notable that expatriates are like any other professionals and are unlikely to be engaged in their countries of origin without the appropriate incentives. Policies or practices that assume that these individuals owe something to their countries of origin are unlikely to work.

Considerable effort needs to be put in fostering atmosphere of trust between the ex-patriates and local communities. In addition, working from a common objective is critical, as illustrated in the case of the reconstruction of Somaliland. In this inspirational example, those involved in the Somaliland diaspora were able to invoke their competence, networks, and access to capital to establish the University of Hargeisa that has already played a critical role in building the human resource base needed for economic development. The achievement is even more illustrative when one considers the fact that the university was built after the collapse of Somalia.[30] Similar efforts involving the Somali diaspora in collaboration with King's College Hospital in London have contributed significantly to the health care sector in Somaliland.[31] There are important lessons in this case that can inform the rest of Africa. The initial departure of nations to acquire knowledge and skills in other countries represents a process of upgrading their skills and knowledge through further training. But returning home without adequate opportunities to deploy the knowledge earned may represent the ultimate brain drain. A study of Sri Lankan scientists in diaspora has shown that further studies "was the major reason for emigration, followed by better career prospects. Engineering was the most common specialization, followed by chemistry, agricultural sciences and microbiology/ biotechnology/molecular biology. If their demands are adequately met, the majority of the expatriates were willing to return to Sri Lanka."[32]

Strengthening Science, Technology, and Engineering Diplomacy

The area of science, technology, and engineering diplomacy has become a critical aspect of international relations.[33] Science is gaining in prominence as a tool fostering cooperation and resolving disputes among nations.[34] Much of the leadership is provided by industrialized countries. For example, the United States has launched a program of science envoys which is adding a new dimension to U.S. foreign policy.[35] This diplomatic innovation is likely to raise awareness of the importance of science, technology, and engineering in African countries.

Ministries of foreign affairs have a responsibility in promoting international technology cooperation and forging strategic alliances. To effectively carry out this mandate, these ministries need to strengthen their internal capability in science, technology, and innovation. To this end, they will need to create offices dealing specifically with science, technology, and engineering, working in close cooperation with other relevant ministries, industry, academia, and civil society. Such offices could also be responsible for engaging and coordinating expatriates in Africa's technology development programs.

There has been growing uncertainty over the viability of traditional development cooperation models. This has inspired the emergence of new technology alliances involving the more advanced developing countries.[36] For example, India, Brazil, and South Africa have launched a technology alliance that will focus on finding solutions to agricultural, health, and environmental challenges. In addition, more developing countries are entering into bilateral partnerships to develop new technologies. Individual countries such as China and Brazil are also starting to forge separate technology-related alliances with African countries. Brazil, for example, is increasing its cooperation with African countries in agriculture and other fields.[37] In addition to establishing a branch of the Brazilian Agricultural Research Corporation (EMBRAPA) in Ghana, the country has also created a tropical agricultural research institute at home to foster cooperation with African countries.

Significant experiments are under way around the world to make effective use of citizens with scientific expertise who are working abroad. The UK consulate in Boston is engaged in a truly pioneering effort to advance science, technology, and engineering diplomacy. Unlike other consulates dealing with regular visa and citizenship issues, the consulate is devoted to promoting science, technology, and engineering cooperation between the UK and the United States while also addressing major global challenges such as climate change and international conflict.

In addition to Harvard University and MIT, the Boston area is home to more than 60 other universities and colleges, making it the de facto intellectual capital of the world. Switzerland has also converted part of its consulate in Boston into a focal point for interactions between Swiss experts in the United States and their counterparts at home. Swissnex was created in recognition of the importance of having liaisons in the area, which many consider the world's leading knowledge center, especially in the life sciences. These developments are changing the way governments envision the traditional role of science attachés, with many giving them more strategic roles.[38]

In another innovative example, the National University of Singapore has established a college at the University of Pennsylvania to focus on biotechnology and entrepreneurship. The complementary Singapore-Philadelphia Innovators' Network (SPIN) serves as a channel and link for entrepreneurs, investors, and advisers in the Greater Philadelphia region and Singapore. The organization seeks to create opportunities for international collaboration and partnerships in the area.

India, on the other hand, has introduced changes in its immigration policy, targeting its citizens working abroad in scientific fields to strengthen their participation in development at home. Such approaches can be adopted by other developing countries, where the need to forge international technology partnerships may be even higher, provided there are institutional mechanisms to facilitate such engagements.[39] The old-fashioned metaphor of the "brain drain" should to be replaced by a new view of "global knowledge flows."[40]

But even more important is the emerging interest among industrialized countries to reshape their development cooperation strategies to reflect the role of science, technology, and innovation in development. The UK Department for International Development (DFID) took the lead in appointing a chief scientist to help provide advice to the government on the role of innovation in international development, a decision that was later emulated by USAID.[41] Japan has launched a program on science and technology diplomacy that seeks to foster cooperation with developing countries on the basis of its scientific and technological capabilities.[42] Similarly, the United States has initiated efforts to place science, technology, and innovation at the center of its development cooperation activities.[43] The initiative will be implemented through USAID as part of the larger science and technology diplomacy agenda of the U.S. government.[44]

South Korea is another industrialized country that is considering adopting a science and innovation approach to development cooperation. These trends might inspire previous champions of development such as Sweden to consider revamping their cooperation programs. These efforts are going to be reinforced by the rise of new development cooperation models in emerging economies such India, Brazil, and China. India is already using its strength in space science to partner with African countries. Brazil, on the other hand, positioning itself as a leading player in agricultural cooperation with African countries, is seeking to expand the activities of the Brazilian Development Cooperation Agency.

China's cooperation with Africa is increasingly placing emphasis on science, technology, and engineering. It is a partner in 100 joint demonstration projects and postdoctoral fellowships, which include donations of nearly US$22,000 worth of scientific equipment. China has offered to build 50 schools train 1,500 teachers and principals as well as train 20,000 professionals by 2012. The country will increase its demonstration centers in Africa to 20, send 50 technical teams to the continent, and train 2,000 African agricultural personnel. Admittedly, these numbers are modest given the magnitude of the challenge, but they show a shift toward using science, technology, and engineering as tools for development cooperation.[45]

Harmonization of Regional Integration Efforts

When the heads of state and government of the Common Market for Eastern and Southern Africa (COMESA), the East African Community (EAC), and the Southern African Development Community (SADC) met in Kampala on October 22, 2008, they conveyed in their communiqué a palpable sense of urgency in calling for the establishment of a single free trade area covering the 26 countries of COMESA, EAC, and SADC. These are 26 of the 54 countries that make up the continent of Africa. The political leaders requested the secretariats of the three organizations to prepare all the legal documents necessary for establishing the single free trade area (FTA) and to clearly identify the steps required—paragraph 14 of the communiqué. In November 2009 the chief executives of the three secretariats cleared the documents for transmission to the member states for consideration in preparing for the next meeting of the Tripartite Summit. The main document is the draft agreement establishing the Tripartite Free Trade, with its 14 annexes covering various complementary areas that are necessary for effective functioning of a regional market. There is a report explaining the approach and the modalities. The main proposal is to establish the FTA on a tariff-free, quota-free, exemption-free basis by simply combining the existing FTAs of COMESA, EAC, and SADC. It is expected that by 2012, none of these FTAs will have any exemptions or sensitive lists. However, there is a possibility that a few countries might wish to consider maintaining a few sensitive products in trading with some big partners, and for this reason, provision has been made for the possibility of a country requesting permission to maintain some sensitive products for a specified period of time.

To have an effective tripartite FTA, various complementary areas have been included. The FTA will cover promotion of customs cooperation and trade facilitation; harmonization and coordination of industrial and health standards; combating of unfair trade practices and import surges; use of peaceful and agreed dispute settlement mechanisms; application of simple and straightforward rules of origin that recognize inland transport costs as part of the value added in production; and relaxation of restrictions on movement of businesspersons, taking into account certain sensitivities.

It will also seek to liberalize certain priority service sectors on the basis of existing programs; promote value addition and transformation of the region into a knowledge-based economy through a balanced use of intellectual property rights and information and communications technology; and develop the cultural industries. The tripartite FTA will be underpinned by robust infrastructure programs designed to consolidate the regional market through interconnectivity (facilitated, for instance, by all modes of transport and telecommunications) and to promote competitiveness (for instance, through adequate supplies of energy).

Regarding the steps required, or the road map, the proposal is that there should be a preparatory period for consultations at the national, regional, and tripartite level from early 2010 to June 2011. Member states will use this period to carefully work out the legal and institutional framework for the single FTA using the draft documents as a basis. It is expected that each organization will discuss the tripartite documents, and, through the tripartite meetings at various levels, will deliberate and reach concrete recommendations. By June 2011, there should be a finalized agreement establishing the Tripartite FTA, ready for signature in July 2011. When signed, member states will have about six months (up to December 2011) to finalize their domestic processes for approving the agreement (for instance, ratification) and for establishing the required institutions and adopting the relevant customs and other documentation and instruments. It is proposed that once this process ends, the Tripartite FTA should be launched in January 2012. Throughout the preparatory period, strong sensitization programs will be mounted for the public and private sectors and all stakeholders including parliamentarians, business community, teaching institutions, civil society, and development partners.

The main benefit of the Tripartite FTA is that it will be a much larger market, with a single economic space, than any one of the three regional economic communities and as such will be more attractive to investment and large-scale production. Estimates are that exports among the 26 tripartite countries increased from US$7 billion in 2000 to US$27 billion in 2008, and imports grew from US$9 billion in 2000 to US$32 billion in 2008. This phenomenal increase was in large measure spurred by the free trade area initiatives of the three organizations. Strong trade performance, when well designed—for instance, by promoting small and medium-scale enterprises that produce goods or services—can assist the achievement of the core objectives of eradicating poverty and hunger, promoting social justice and public health, and supporting all-round human development. Besides, the tripartite economic space will help to address some current challenges resulting from multiple membership by advancing the ongoing harmonization and coordination initiatives of the three organizations to achieve convergence of programs and activities, and in this way will greatly contribute to the continental integration process. And as they say, the more we trade with each other, the less likely we are to engage in war, for our swords will be plowshares.

Harmonization of Regulations

The need to enhance the use of science, technology, and engineering in development comes with new risks. Africa has not had a favorable history with new technologies. Much of its history has been associated with the use of technology as tools of domination or extraction.[46] The general mood of skepticism toward technology and the long history of exclusion created a political atmosphere that focused excessively on the risks of new technologies. This outlook has been changing quite radically as Africa enters a new era in which the benefits of new technologies to society are widely evident. These trends are reinforced by political shifts that encourage great social inclusion.[47] It is therefore important to examine the management of technological risks in the wider social context even if the risk assessment tools that are applied are technical.

The risks associated with new technologies need to be reviewed on a case-by-case basis and should be compared with base scenarios, many of which would include risks of their own. In other words, deciding not to adopt new technologies may only compound the risks associated with the status quo. Such an approach would make risk management a knowledge-based process. This would in turn limit the impact of popular tendencies that prejudge the risks of technologies based on their ownership or newness.

Ownership and newness may have implications for technological risks but they are not the only factors that need to be considered. Fundamentally, decisions on technological risks should take into account the impacts of incumbent technologies or the absence of any technological solutions to problems.

One of the challenges facing African countries is the burden of managing technological risks through highly fragmented systems in contiguous countries. The growing integration of African countries through the RECs offers opportunities to rationalize and harmonize their regulatory activities related to agricultural innovation.[48]

This is already happening in the medical sector. The African Medicines Regulatory Harmonization (AMRH) initiative was established to assist African countries and regions to respond to the challenges posed by medicine registration, as an important but neglected area of medicine access. It seeks to support African Regional Economic Communities and countries in harmonizing medicine registration.

COMESA, in collaboration with the Association for Strengthening Agricultural Research in Eastern and Central Africa (ASARECA) and other implementing partners, has engaged in the development of regionally harmonized policies and guidelines through the Regional Agricultural Biotechnology and Biosafety Policy in Eastern and Southern Africa (RABESA) initiative since 2003. The COMESA harmonization agenda—now implemented through its specialized agency the Alliance for Commodity Trade in Eastern and Central Africa (ACTESA)—was initiated to provide mechanisms for wise and responsible use of genetically modified organisms (GMOs) in commercial planting, trade, and emergency food assistance.

The draft policies and guidelines are expected to be endorsed soon by its policy organ. The guidelines seek the establishment of a COMESA biosafety and centralized GMO risk assessment desk, with standard operating procedures. ACTESA has established a biotechnology program to coordinate biotechnology- and biosafety-related activities and provide guidance in the region. The COMESA panel of experts on biotechnology has been established to provide technical assistance in policy formulation and GMO risk assessment.

The establishment of the World Trade Organization (WTO) in 1995 and the coming into force of a multilateral trading system backed by legally binding trade agreements placed fresh challenges on WTO member states, particularly African countries that were already struggling with trade liberalization and globalization that encouraged the free movement of humans, animals, food, and agricultural products across borders.

Specifically, the agreement on sanitary and phytosanitary measures (SPS) required governments to apply international standards and establish science-based legal and regulatory systems to manage health and environmental risks associated with food and agricultural products. International standards to derive legislation and regulations for SPS management include risk analysis and use of the HACCP) system. The production of food products involves complex production and processing methods. For many African governments, this required new skills to conduct scientific risk analysis along the food chain as food products interact with plant and animal diseases, pests, biological hazards such as pathogenic microorganisms, and naturally occurring hazards such as aflatoxins. Risk analysis requires scientific skills and innovation that go beyond conventional training; if not done properly, this risk analysis will result in weak SPS systems, which in turn may result in nontariff barriers in regional and global markets.

COMESA, within its mandate of regional economic integration, recognizes the need to support member states in resolving non-tariff barriers that constrain markets and stifle the integration of food products into regional and global value chains, as an innovative strategy to promote market access to regional and international trade.

In 2005, COMESA commissioned a project to support member states in their efforts to address SPS barriers by improving and harmonizing SPS measures and food safety systems among member states. Country-level focal points were provided to facilitate and initiate the harmonization process and training was provided on how to establish national and regional surveillance systems. A second step involved the establishment of regional reference laboratories for food safety and animal and plant health. Initial training was provided for key laboratory personnel, and guidelines for regional harmonization of SPS measures and the use of reference laboratories were developed. COMESA will now present the harmonization guidelines to its Technical Committee on Agriculture for regional approval and adoption.

The primary aim of harmonizing medicine registration is to improve public health, by increasing timely access to safe and efficacious medicines of good quality for the treatment of priority diseases. Access will be improved by reducing the time it takes for priority essential medicines to be registered in-country (including the time needed for industry to prepare their registration application or dossier) and so potentially the time taken for essential therapies to reach patients in need (depending on funding and distribution mechanisms). This will include capacity building to ensure transparent, efficient, and competent regulatory activities (assessment of registration dossiers and related inspections) that are able to assure the quality, safety, and efficacy of registered medicines. The AMRH initiative approach seeks to support the RECs and countries to harmonize medicine registration using existing political structures and building on existing plans and commitments.

The AMRH initiative was the outcome of NEPAD) and the Pan-African Parliament, which was hosted in collaboration with consortium partners and attracted representation from nine of the continent's Regional Economic Communities (RECs) and over 40 national medicine regulatory authorities (NMRA). This provided a strong endorsement for the consensus plan that emerged and hence the approach that RECs and NMRAs are now adopting.

The AU Summit approved the Pharmaceutical Manufacturing Plan for Africa in 2007, which specifically recognizes the need for African countries to strengthen their medicine regulatory systems by pooling their resources to achieve public health policy priorities.

Such systems are vital to assuring the quality, safety, and efficacy of locally manufactured products and their positive contribution to public health. Moreover, the success of domestic production will partly depend on intra-regional and intra-continental trade to create viable market sizes. Currently, trade in pharmaceuticals is hampered by disparate regulatory systems, which create technical barriers to the free movement of products manufactured in Africa (and beyond)—and has negative consequences for timely patient access to high-quality essential medicines.

Several RECs have already supported harmonization of medicine registration by developing common pharmaceutical policies and operational plans—backed by high-level political commitments and mandates. For example, in east Africa under the provisions of Chapter 21 (Article 118) of the EAC treaty, medicine registration harmonization is an explicit policy priority. Likewise, in southern Africa, ministers of health have approved the SADC Pharmaceuticals Business Plan, with explicit goals to harmonize medicine registration.

However, implementation of these policies and plans has suffered from a lack of financial and technical resources and has not progressed significantly. Moreover, RECs continue to work largely in isolation. Coordination is needed to avoid duplication of effort and ensure consistent approaches, especially given that more than three-quarters of African countries belong to two or more RECs. Following commencement of the AMRH initiative, several RECs submitted summary project proposals describing their high-level plans for harmonization of medicine registration.

Following their review, the consortium actively partnered with four REC groupings (EAC, Economic Community of Central African States, ECOWAS, the West African Monetary Union and SADC) to support them in strengthening and moderately expanding their proposals. In the first instance, this involved written feedback from the consortium followed by a visit from a NEPAD delegation to explain the consortium's feedback and agree on a timeline for next steps. Given the importance of NMRA consultations and ownership, NEPAD has also made some funding available for RECs and their constituent NMRAs to jointly ensure that their proposals reflect their shared vision for harmonized medicine registration.

Conclusion

Promoting a growth-oriented agenda will entail adjustments in the structure and functions of government. More fundamentally, issues related to science, technology, and innovation will need to be addressed in an integrated way at the highest level possible in government. Bringing science, technology, and engineering to the center of Africa's economic renewal will require more than just political commitment; it will take executive leadership. This challenge requires concept champions who in this case will be heads of state spearheading the task of shaping their economic policies around science, technology, and innovation.

So far, most African countries have not developed national policies that demonstrate a sense of focus to help channel emerging technologies into solving developmental problems. They still rely on generic strategies dealing with "poverty alleviation," without serious consideration of the sources of economic growth. There are signs of hope though. NEPAD's Ministerial Forum on Science and Technology played a key role in raising awareness among African's leaders of the role of science, technology, and engineering in economic growth.

An illustration of this effort is the decision of the African Union (AU) and NEPAD to set up a high level African Panel on Modern Biotechnology to advise the AU, its member states, and its various organs on current and emerging issues associated with the development and use of biotechnology. The panel's goal is to provide the AU and NEPAD with independent and strategic advice on biotechnology and its implications for agriculture, health, and the environment. It focuses on intra-regional and international regulation of the development and application of genetic modification and its products.

Regarding food security in particular, Africa's RECs have tried to develop regional policies and programs to allow member states to work collectively. ECOWAS and COMESA, for instance, building on CAADP, have elaborated regional compacts to guide member states in formulating their national CAADP compacts. This comes at a time when experience from the six COMESA national CAADP compacts so far concluded show that there are key cross-border challenges that need a regional approach. Twenty-six African nations had signed CAADP compacts by June 2010, compounding the need for regional strategies.

Thus, there is a need for a larger regional market to support investment in agricultural products and harmonization of standards across the region. This will help address challenges such sanitary and phytosanitary measures that affect the quality and marketability of agricultural products; management of transboundary resources such as water bodies and forests; building of regional infrastructure; promotion of collaborative research; monitoring of key commitments of member states, particularly the one on earmarking 10% of the national budget for the agriculture sector. A key pillar of CAADP relates to agricultural research and innovation. Africa's RECs have a critical role to play under this pillar, through supporting regional research networks and prioritizing agricultural research in regional policies. Experience sharing at the regional level, and the resulting research communities, will greatly enrich individual research.

8

CONCLUSIONS AND THE WAY AHEAD

A new economic vision for Africa's agricultural transformation—articulated at the highest level of government through Africa's Regional Economic Communities (RECs)—should be guided by new conceptual frameworks that define the continent as a learning society. This shift will entail placing policy emphasis on emerging opportunities such as renewing infrastructure, building human capabilities, stimulating agribusiness development, and increasing participation in the global economy. It also requires an appreciation of emerging challenges such as climate change and how they might influence current and future economic strategies.

Climate Change, Agriculture, and Economy

As Africa prepares to address its agricultural challenges, it is now confronted with new threats arising from climate change. Agricultural innovation will now have to be done in the context of a more uncertain world in which activities such as plant and animal breeding will need to be anticipatory.[1] According to the World Bank, warming "of 2°C could result in a 4 to 5 percent permanent reduction in annual income per capita in Africa and South Asia, as opposed to minimal losses in high-income countries and a global average GDP loss of about 1 percent. These losses would be driven by impacts in agriculture, a sector important to the economies of both Africa and South Asia."[2] Sub-Saharan Africa is dominated by fragile ecosystems. Nearly 75% of its surface area is dry land or desert. This makes the continent highly vulnerable to droughts and floods. Traditional cultures cope with such fragility through migration. But such migration has now become a source of insecurity in parts of Africa. Long-term responses will require changes in agricultural production systems.[3]

The continent's economies are also highly dependent on natural resources. Nearly 80% of Africa's energy comes from biomass and over 30% of its GDP comes from rain-fed agriculture, which supports 70% of the population. Stress is already being felt in critical resources such as water supply. Today, 20 African countries experience severe water scarcity and another 12 will be added in the next 25 years. Economic growth in regional hubs is now being curtailed by water shortages.

The drying up of Lake Chad (shared by Nigeria, Chad, Cameroon, and Niger) is a grim reminder that rapid ecological change can undermine the pursuit for prosperity. The lake's area has decreased by 80% over the last 30 years, with catastrophic impacts to local communities. Uncertainty over water supply affects decisions in other areas such as hydropower, agriculture, urban development, and overall land-use planning. This is happening at a time when Africa needs to switch to low-carbon energy sources.

Technological innovation will be essential for enabling agriculture to adapt to a different climate. Meeting the dual challenges of expanding prosperity and adapting to climate change will require greater investment in the generation and diffusion of new technologies. Basic inputs such as provision of meteorological data could help farmers to adapt to climate change by choosing optimal planting dates.[4] The task ahead for policy makers will be to design climate-smart innovation systems that shift economies toward low-carbon pathways. Economic development is an evolutionary process that involves adaptation to changing economic environments. Technological innovation is implicitly recognized as a key aspect of adaptation to climate change. For example, the Intergovernmental Panel on Climate Change (IPCC) defines adaptation as "Adjustment in natural or human systems in response to actual or expected climatic stimuli or their effects, which moderates harm or exploits beneficial opportunities."[5] It views the requisite adaptive capacity as the ability "to moderate potential damages, to take advantage of opportunities, or to cope with the consequences."[6] Technological innovation is used in society in a congruent way to respond to economic uncertainties. What is therefore needed is to develop analytical and operational frameworks that would make it easier to incorporate adaptation to climate change in innovation strategies used to expand prosperity.

Innovation systems are understood to mean the interactive process involving key actors in government, academia, industry, and civil society to produce and diffuse economically useful knowledge into the economy. The key elements of innovation include the generation of a variety of avenues, their selection by the market environment, and the emergence of robust socioeconomic systems. This concept can be applied to adaption to climate change in five critical areas: managing natural resources; designing physical infrastructure; building human capital, especially in the technical fields; fostering entrepreneurial activities; and governing adaptation as a process of innovation.

Economic development is largely a process by which knowledge is applied to convert natural resources into goods and services. The conservation of nature's variety is therefore a critical aspect of leaving options open for future development. Ideas such as "sustainable development" have captured the importance of incorporating the needs of future generations into our actions. Adaptive strategies will therefore need to start with improved understanding of the natural resource base. Recent advances in earth observation and related geospatial science and technology have considerably increased the capacity of society to improve its capabilities for natural resource management. But improved understanding is only the first step.

The anticipated disruptive nature of climate change will demand increased access to diverse natural assets such as genetic resources for use in agriculture, forestry, aquaculture, and other productive activities. For example, the anticipated changes in the growing season of various crops will require intensified crop breeding.[7] But such breeding programs will presuppose not only knowledge of existing practices but also the conservation of a wider pool of genetic resources of existing crops and breeds and their wild relatives to cope with shifts in agricultural production potential.[8] This can be done through measures such as seed banks, zoos, and protected areas. Large parts of Africa may have to switch from crop production to livestock breeding.[9] Others may also have to change from cultivating cereals to growing fruits and vegetables as projected in other regions of the world.[10] Other measures will include developing migration corridors to facilitate ecosystem integrity and protect human health—through surveillance and early warning systems.

Such conservation efforts will also require innovation in regional institutional coordination, expanded perspectives of space and time, and the incorporation of climate change scenarios in economic development strategies.[11] Building robust economies requires the conservation of nature's variety. These efforts will need to be accompanied by greater investment in the generation of knowledge associated with natural resources. Advances in information and communication capabilities will help the international community to collect, store, and exchange local knowledge in ways that were not possible in the past. The sequencing of genomes provides added capacity for selective breeding of crops and livestock suited to diverse ecologies. Technological advancement is therefore helping to augment nature's diversity and expand adaptive capabilities.

Climate change is likely to affect existing infrastructure in ways that are not easy to predict. For example, road networks and energy sources in low-lying areas are likely to be affected by sea level rise. A recent study of Tangier Bay in Morocco projects that sea level rise will have significant impact on the region's infrastructure facilities such as coastline protection, the port, railway lines, and the industrial base in general.[12]

Studies of future disruptions in transportation systems reveal great uncertainties in impact depending on geographical location.[13] These uncertainties are likely to influence not only investment decisions but also the design of transportation systems. Similarly, uncertainty over water supply is emerging as a major concern demanding not only integrated management strategies but also improved use of water-related technologies.

Other measures include the need to enhance water supply— such as linking reservoirs, building new holding capacity in reservoirs, and injecting early snowmelt into groundwater reservoirs. Similarly, coastal areas need to be protected with natural vegetation or seawalls. In effect, greater technical knowledge and engineering capabilities will need to be marshaled to design future infrastructure in light of climate change.[14] This includes the use of new materials arising from advances in fields such as nanotechnology.

Protecting human populations from the risks of climate change should be one of the first steps in seeking to adapt to climate change. Concern over human health can compound the sense of uncertainty and undermine other adaptive capabilities. Indeed, the first step in building resilience is to protect human populations against disease.[15] Many of the responses needed to adapt health systems to climate change will involve practical options that rely on existing knowledge.[16]

Others, however, will require the generation of new knowledge. Advances in fields such as genomics are making it possible to design new diagnostic tools that can be used to detect the emergence of new infectious diseases. These tools, combined with advances in communications technologies, can be used to detect emerging trends in health and provide health workers with early opportunities to intervene. Furthermore, convergence in technological systems is transforming the medical field. For example, the advent of hand-held diagnostic devices and video-mediated consultation are expanding the prospects of telemedicine and making it easier for isolated communities to be connected to the global health infrastructure.[17] Personalized diagnostics is also becoming a reality.[18]

Adapting to climate change will require significant upgrading of the knowledge base of society. Past failure to adapt from incidences of drought is partly explained by the lack of the necessary technical knowledge needed to identify trends and design responses.[19] The role of technical education in economic development is becoming increasingly obvious. Similarly, responding to the challenges of climate will require considerable investment in the use of technical knowledge at all levels in society.

One of the most interesting trends is the recognition of the role of universities as agents of regional economic renewal.[20] Knowledge generated in centralized urban universities is not readily transferred to regions within countries. As a result, there is growing interest in decentralizing the university system itself.[21] The decentralization of technical knowledge to a variety of local institutions will play a key role in enhancing local innovation systems that can help to spread prosperity through climate-smart strategies.

The ability to adapt to climate change will not come without expertise. But expertise is not sufficient unless it is used to identify, assess, and take advantage of emerging opportunities through the creation of new institutions or the upgrading of existing ones. Such entrepreneurial acts are essential both for economic development and adaptation to climate change. Economic diversification is critical in strengthening the capacity of local communities to adapt to climate change.

For example, research on artisan fisheries has shown that the poorest people are not usually the ones who find it hardest to adapt to environmental shocks. It is often those who have become locked in overly specialized fishery practices.[22] Technological innovation aimed at promoting diversification of entrepreneurial activities would not only help to improve economic welfare, but it would also help enhance the adaptive capabilities of local communities. But such diversification will need to be complemented by other measures such as flexibility, reciprocity, redundancy, and buffer stocks.[23]

Promoting prosperity and creating robust economies that can adapt to climate change should be a central concern of leaders around the world. Political turmoil in parts of Africa is linked to recent climate events.[24] The implications of climate change for governance, especially in fragile states, has yet to receive attention.[25] Governments will need to give priority to adaption to climate change as part of their economic development strategies. But they will also need to adopt approaches that empower local communities to strengthen their adaptive capabilities. Traditional governance practices such as participation will need to be complemented by additional measures that enhance social capital.[26]

The importance of technological innovation in adaptation strategies needs to be reflected in economic governance strategies at all levels. It appears easier to reflect these considerations in national economic policies. However, similar approaches also need to be integrated into global climate governance strategies, especially through the adoption of technology-oriented agreements.[27]

On the whole, an innovation-oriented approach to climate change adaptation will need to focus largely on expanding the adaptive capacity of society though the conservation of nature's variety, construction of robust infrastructure, enhancement of human capabilities, and promotion of entrepreneurship. Fundamentally, the ability to adapt to climate change will possibly be the greatest test of our capacity for social learning. Regional integration will provide greater flexibility and geographical space for such learning. Furthermore, promoting local innovation as part of regional strategies will contribute to the emergence of more integrated farming systems.[28]

Throughout, this book has highlighted the role that RECs can have as a collective framework for harnessing national initiatives and sharing best practices drawn from the region and beyond. Africa's RECs, as well as the African Union at the continental level, have programs for food security, and for science, technology, and engineering. The challenge, as highlighted, relates to putting existing knowledge within the region and beyond to the service of the people of Africa on the ground, through clear political and intellectual leadership and an effective role for innovators. Further, there is a challenge of how best to utilize the existing regional policy making and monitoring and evaluation structures in promoting innovation and tackling the challenges of food security.

The current global economic crisis and rising food prices are forcing the international community to review their outlook for human welfare and prosperity. Much of the current concern on how to foster development and prosperity in Africa reflects the consequences of recent neglect of sustainable agriculture and infrastructure as drivers of development. Sustainable agriculture has, through the ages, served as the driving force behind national development. In fact, it has been a historical practice to use returns from investment in sustainable agriculture to stimulate industrial development. Restoring it to its right place in the development process will require world leaders to take a number of bold steps.

Science and innovation have always been the key forces behind agricultural growth in particular and economic transformation in general. More specifically, the ability to add value to agricultural produce via the application of scientific knowledge to entrepreneurial activities stands out as one of the most important lessons of economic history. Reshaping sustainable agriculture as a dynamic, innovative, and rewarding sector in Africa will require world leaders to launch new initiatives that include the following strategic elements.

Bold leadership driven by heads of state in Africa, supported by those of developed and emerging economies, is needed to recognize the real value of sustainable agriculture in the economy of Africa. High-level leadership is essential for establishing national visions for sustainable agriculture and rural development, championing of specific missions for lifting productivity and nutritional levels with quantifiable targets, and the engagement of cross-sectoral ministries in what is a multi-sector process.

Sustainable agriculture needs to be recognized as a knowledge-intensive productive sector that is mainly carried out in the informal private economy. The agricultural innovation system has to link the public and private sectors and create close interactions between government, academia, business, and civil society. Reforms will need to be introduced in knowledge-based institutions to integrate research, university teaching, farmers' extension, and professional training, and bring them into direct involvement with the production and commercialization of products.

Policies have to urgently address affordable access to communication services for people to use in their everyday lives, as well as broadband Internet connectivity for centers of learning such as universities and technical colleges. This is vital to access knowledge and trigger local innovations, boosting rural development beyond sustainable agriculture. It is an investment with high returns. Improving rural productivity also requires significant investments in basic infrastructure including transportation, rural energy, and irrigation. There will be little progress without such foundational investments.

Fostering entrepreneurship and facilitating private sector development has to be highest on the agenda to promote the autonomy and support needed to translate opportunity into prosperity. This has to be seen as an investment in itself, with carefully tailored incentives and risk-sharing approaches supported by government.

Entrepreneurial Leadership

It is not enough for governments to simply reduce the cost of doing business. Fostering agricultural renewal will require governments to function as active facilitators of technological learning. Government actions will need to reflect the entrepreneurial character of the farming community; they too will need to be entrepreneurial.[29] Leadership will also need to be entrepreneurial in character.[30] Moreover, addressing the challenge will require governments to adopt a mission-oriented approach, setting key targets and providing support to farmers to help them meet quantifiable goals. A mission-oriented approach will require greater reliance on executive coordination of diverse departmental activities.

Fostering economic renewal and prosperity in Africa will entail adjustments in the structure and functions of government. More fundamentally, issues related to agricultural innovation must be addressed in an integrated way at the highest possible levels in government. There is therefore a need to strengthen the capacity of presidential offices to integrate science, technology, and innovation in all sustainable agriculture-related aspects of government. Moreover, such offices will also need to play a greater role in fostering interactions between government, business, academia, and civil society. This task requires champions.

One of the key aspects of executive direction is the extent to which leaders are informed about the role of science and innovation in agricultural development. Systematic advice on science and innovation must be included routinely in policy making.[31] Such advisers must have access to credible scientific or technical information drawing from a diversity of sources including scientific and engineering academies. In fact, the magnitude of the challenge for regions like Africa is so great that a case could be made for new academies dedicated to agricultural science, technology, and innovation.

Science, technology, and engineering diplomacy has become a critical aspect of international relations. Ministries of foreign affairs in African countries have a responsibility to promote international technology cooperation and forge strategic alliances on issues related to sustainable agriculture. To effectively carry out this task, foreign ministries need to strengthen their internal capability in science and innovation.

Toward a New Regional Economic Vision

Contemporary history informs us that the main explanation for the success of the industrialized countries lies in their ability to learn how to improve performance in a diversity of social, economic, and political fields. In other words, the key to their success was their focus on practical knowledge and the associated improvements in skills needed to solve problems. They put a premium on learning based on historical experiences.[32]

One of the most reassuring aspects of a learner's strategy is that every generation receives a legacy of knowledge that it can harness for its own use. Every generation blends the new and the old and thereby charts its own development path, making debates about innovation and tradition irrelevant. Furthermore, discussions on the impact of intellectual property rights take on a new meaning if one considers the fact that the further away you are from the frontier of research, the larger is your legacy of technical knowledge. The challenge therefore is for Africa to think of research in adaptive terms, and not simply focus on how to reach parity with the technological front-runners. Understanding the factors that help countries to harness available knowledge is critical to economic transformation.

The advancement of information technology and its rapid diffusion in recent years could not have happened without basic telecommunication infrastructure. In addition, electronic information systems, which rely on telecommunications infrastructure, account for a substantial proportion of production and distribution activities in the secondary and tertiary sectors of the economy. It should also be noted that the poor state of Africa's telecommunications infrastructure has hindered the capacity of the region to make use of advances in fields such as geographical information sciences in sustainable development.

The emphasis on knowledge is guided by the view that economic transformation is a process of continuous improvement in productive activities. In other words, government policy should be aimed at enhancing performance, starting with critical fields such as agriculture, while recognizing interdisciplinary linkages.

This type of improvement indicates a society's capacity to adapt to change through learning. It is through continuous improvement that nations transform their economies and achieve higher levels of performance. Using this framework, with government functioning as a facilitator for economic learning, agribusiness enterprises will become the locus of learning, and knowledge will be the currency of change.

Some African countries already possess the key institutional components they need to become players in the knowledge economy. The emphasis, therefore, should be on realigning the existing structures and creating necessary new ones where they do not exist and promoting interactions between key players in the economy. More specifically, the separation between government, industry, and academia stands out as one of the main sources of inertia and waste in Africa's knowledge-based institutions.[33] The challenge is not simply creating institutions, but creating systems of innovation in which emphasis is placed on economic learning through interactions between actors in the society.

A key role of Africa's RECs is to provide the regional framework for all stakeholders to act in a coordinated manner, share best practices, encourage peer review of achievements and setbacks by key players, and pool resources for the greater good of the region and Africa at large. The policy organs of the RECs, including the presidents and sectoral ministers, provide appropriate frameworks for the public and private sector to formulate innovative policies; and given the multidisciplinary and multi-sectoral nature of the initiatives, the higher policy organs at the level of heads of state and government, and at the level of joint ministerial meetings, provide a unique role for the RECs as vehicles for promoting regional collaboration and for the elaboration and implementation of key policy initiatives.

Africa has visions for socioeconomic development at the national, regional, and continental level. Science, technology, engineering, and innovation are critical pillars of any socioeconomic development vision in our time. At the three levels, the visions do not coherently interact because the continental policies are not necessarily coordinated with the policies the member states adopt and implement in the context of the RECs, and national policy making is at times totally divorced from the regional and continental processes and frameworks.

However, there are case studies of how some RECs have tried to address this dilemma, which could constitute best practices for implementation of regional policies at the national level and for elaboration of regional policies on the basis of practical realities in the member states. In the East African Community (EAC), each member state has agreed to establish a dedicated full-scale ministry responsible for EAC affairs. This means that EAC affairs are organically integrated into the national government structure of the member states. There is need for a coherent approach to formulation and implementation of regional policies at the national level, drawing on the collective wisdom and clout that RECs provide in tackling key national and regional challenges, particularly those related to the rapid socioeconomic transformation of Africa.

APPENDIX I

REGIONAL ECONOMIC COMMUNITIES (RECS)

The modern evolution of Africa's economic governance can be traced to the 1980 Lagos Plan of Action for the Development of Africa and the 1991 Abuja Treaty that established the African Economic Community (AEC). The treaty envisaged Regional Economic Communities (RECs) as the building blocks for the AEC. It called on member states to strengthen existing RECs and provided timeframes for creating new ones where they did not exist. Eight RECs have been designated by the African Union as the base for Africa's economic integration.

Arab Maghreb Union (AMU)

The first Conference of Maghreb Economic Ministers in Tunis in 1964 established the Conseil Permanent Consultatif du Maghreb between Algeria, Libya, Morocco, and Tunisia. Its aim was to coordinate and harmonize the development plans of the four countries, foster intraregional trade, and coordinate relations with the European Union. However, the plans never came to fruition. The first Maghreb Summit, held in Algeria in 1988, decided to set up the Maghreb High Commission. In 1989 in Marrakech the Treaty establishing the AMU was signed. The main objectives of AMU are to ensure regional stability and enhance policy coordination and to promote the free movement of goods and services.

The five member countries have a total population of 85 million people and a GDP of US$500 billion. The headquarters of AMU are in Rabat-Agdal, Morocco. Unlike the other RECs, AMU has no relations with the African Economic Community (AEC) and has not yet signed the Protocol on Relations with the AEC. Morocco is not a member of the AU. The AMU, however, been designated by the African Union as a pillar of the AEC.

Common Market for Eastern and Southern Africa (COMESA)

The Common Market for Eastern and Southern Africa was founded in 1993 as a successor to the Preferential Trade Area for Eastern and Southern Africa (PTA), which was established in 1981. Its vision is to "be a fully integrated, internationally competitive regional economic community with high standards of living for all its people ready to merge into an African Economic Community." Its mission is to "achieve sustainable economic and social progress . . . through increased co-operation and integration in all fields of development particularly in trade, customs and monetary affairs, transport, communication and information, technology, industry and energy, gender, agriculture, environment and natural resources." COMESA formally succeeded the PTA in 1994. The establishment of COMESA fulfilled the requirements of the PTA to become a common market. COMESA has 19 member states with a population of 410 million with a total GDP of over US$360 million. It has an annual import bill of about US$152 billion and an export bill of over US$157 billion. COMESA forms a major market place for both internal and external trading. Its headquarters are in Lusaka, Zambia.

Community of Sahel-Saharan States (CEN-SAD)

The Community of Sahel-Saharan States was established in 1998 following the conference of leaders and heads of states held in Tripoli by Libya, Burkina Faso, Mali, Niger, Chad, and Sudan. Twenty-two more countries have joined the community since then. Its goal is to strengthen peace, security, and stability and achieve global economic and social development. CEN-SAD has signed partnership agreements with numerous regional and international organizations to collaborate on political, cultural, economic, and social issues. Two of its main areas of work are security and environmental management, which include its flagship project to create the Great Green Wall of trees across the Sahel. CEN-SAD member states have launched periodic international sporting and cultural festivals. The first CEN-SAD Games were held in Niamey, Niger, in 2009. Thirteen nations competed in Under-20 sports (athletics, basketball, judo, football, handball, table tennis, and traditional wrestling) and six fields of cultural competition (song, traditional creation and inspiration dancing, painting, sculpture, and photography). The second CEN-SAD Games are scheduled to take place in 2011 in N'Djamena, Chad.

East African Community (EAC)

The East African Community was created in 1967. It collapsed in 1977 due to political differences and was revived in 1996 when the Secretariat of the Permanent Tripartite Commission for East African Cooperation was set up at the EAC headquarters in Arusha, Tanzania. In 1997 the East African heads of state started the process of upgrading the agreement that set up the commission into a treaty. In 1999 they signed the treaty reestablishing the East African Community (EAC). The objectives of EAC are to develop policies and programs aimed at widening and deepening cooperation on political, economic, social, and cultural fields, research and technology, defense, security, and legal and judicial affairs. Its members are now Kenya, Uganda, Tanzania, Rwanda, and Burundi with a population of 130 million and a GDP of over US$80 billion. The EAC has an operating customs union and launched its common market in July

2010. Its roadmap includes a common currency and the creation of a single state. In addition to its secretariat, the EAC has a judiciary and a legislature made of representatives from member states.

Economic Community of Central African States (ECCAS)

At a summit meeting in December 1981, the leaders of the Central African Customs and Economic Union agreed in principle to form a wider economic community of Central African states. ECCAS was established on October 18, 1983, by the union members and the members of the Economic Community of the Great Lakes States (Burundi, Rwanda, and the then Zaire) as well as São Tomé and Principe. Angola remained an observer until 1999, when it became a full member. ECCAS aims "to promote and strengthen harmonious cooperation and balanced and self-sustained development in all fields of economic and social activity, particularly in the fields of industry, transport and communications, energy agriculture, natural resources, trade, customs, monetary and financial matters, human resources, tourism, education, further training, culture, science and technology and the movement of persons." ECCAS began functioning in 1985 but has been inactive since 1992 because of financial difficulties (nonpayment of membership fees) and the conflict in the Great Lakes area. Its 11 member states have a total population of 125 million and a GDP of US$180 billion. Its headquarters are in Libreville, Gabon.

Economic Community of West African States (ECOWAS)

The idea for a West African community goes back to President William Tubman of Liberia, who made the call in 1964. An agreement was signed between Côte d'Ivoire, Guinea, Liberia, and Sierra Leone in February 1965, but this came to nothing. In April 1972 Nigerian and Togolese leaders relaunched the idea, drew up proposals, and toured 12 countries, soliciting their plan from July to August 1973. A 1973 meeting in Lomé studied a draft treaty. A treaty setting up the Economic Community of West African States was signed in 1975 by 15 countries. Its mission is to promote economic integration in all fields of economic activity, particularly industry, transport, telecommunications, energy, agriculture, natural resources, commerce, monetary and financial questions, and social and cultural matters. ECOWAS members signed a nonaggression protocol in 1990, building on two earlier agreements of 1978 and 1981. They also signed a Protocol on Mutual Defense Assistance in 1981 that provided for the creation of an Allied Armed Force of the Community. ECOWAS has 15 members, a population of about 270 million and an estimated GDP of US$380 billion. Its head offices are in Abuja, Nigeria.

Intergovernmental Authority for Development (IGAD)

The Intergovernmental Authority on Drought and Development (IGADD) was formed in 1986 with a very narrow mandate to deal with drought and desertification. IGADD later became the accepted vehicle for regional security and political dialogue among its members. In the mid-1990s IGADD decided to transform the organization into a fully-fledged regional political, economic, development, trade, and security entity. In 1996 the Agreement Establishing the Intergovernmental Authority on Development was adopted. Its transition to economic issues is reflected in its first objective, to promote "joint development strategies and gradually harmonize macroeconomic policies and programs in the social, technological and scientific fields." More specifically, IGAD seeks to "harmonize policies with regard to trade, customs, transport, communications, agriculture, and natural resources, and promote free movement of goods, services, and people and the establishment of residence." In recent years IGAD has been working on environmental security, given the links between conflict and natural resources among its members. It has also continued to work on drought and desertification. IGAD has seven members with a total population of 190 million and a GDP of US$230 billion. It is headquartered in Djibouti City, Djibouti.

Southern African Development Community (SADC)

The Southern African Development Community started as Frontline States whose objective was political liberation of Southern Africa. SADC was preceded by the Southern African Development Coordination Conference (SADCC), which was formed in Lusaka, Zambia, in 1980. The concept of a regional economic cooperation in Southern Africa was first discussed at a meeting of the Frontline States foreign ministers in 1979 in Gaborone, Botswana. The meeting led to an international conference in Arusha, Tanzania, which in turn led to the Lusaka Summit held in 1980. The SADC Treaty and Declaration signed in Windhoek, Namibia, transformed SADCC into SADC. The treaty set out to "promote sustainable and equitable economic growth and socio-economic development that will ensure poverty alleviation with the ultimate objective of its eradication, enhance the standard and quality of life of the people of Southern Africa and support the socially disadvantaged through regional integration." One way set out to achieve this goal is to "promote the development, transfer and mastery of technology." The organization's 15 member states have a total population of 240 million and a GDP of US$750 billion. Its head offices are in Gaborone, Botswana.

APPENDIX II

DECISIONS OF THE 2010 COMESA SUMMIT ON SCIENCE AND TECHNOLOGY FOR DEVELOPMENT

Every year the Common Market for Eastern and Southern Africa (COMESA) chooses a theme to guide its activities for regional integration. The theme for the 2010 COMESA Summit held in Swaziland was "Harnessing Science and Technology for Development." The Chairman of COMESA for this year, His Majesty King Mswati III of the Royal Kingdom of Swaziland, stressed the need for concrete initiatives on science, technology, and innovation that can lead to tangible results for the region. In his address to the summit, he underscored the critical importance of science and technology and made some concrete proposals. He proposed the establishment of technology parks, the creation of an Information and Communication Technology (ICT) Training and Skills Development Fund, and the elaboration of a common ICT curriculum for COMESA to introduce young people to ICT at an early age. He undertook to do everything possible to ensure that the science and technology programs are implemented as agreed.

The Council of Ministers, at their meeting, reached concrete decisions, which the summit fully endorsed. The deliberations in Council are set out below.

Report of the Twenty Eighth Meeting of the COMESA Council of Ministers, Ezulwini, Kingdom of Swaziland, 27–28 August 2010

The Council received a video recorded presentation from Calestous Juma, a Kenyan national who is a Professor of Development Practice at Harvard Kennedy School. Building on a paper circulated for the meeting as the background document for the agenda item on harnessing science and technology, the presentation underscored the importance of science and technology for development and provided a historical perspective to cycles of technological revolutions over the years, as well as a critical discussion of contemporary issues that Africa faces in pursuing its development priorities, suggesting concrete ways forward, with examples, on key issues. Copies of the presentation will be made available to delegations.

The presentation made the following recommendations for establishing an institutional framework for harnessing science and technology in COMESA:

- Creating a high-level committee of science, technology and innovation;
- Establishing offices of science, technology and innovation at the highest level of Government in the Member States and at the Secretariat to support the Governments of the Member States and the Secretary General respectively;
- Promoting regional academies of science, technology and engineering;
- Establishing an Innovation Award for outstanding accomplishment; and
- Setting up a professional or graduate school of regional integration.

In considering these recommendations, Council urged Member States to establish this institutional framework at the national level. Furthermore, the Secretariat should have an Advisory Office on Science and Technology.

In addition, the presentation made specific recommendations on harnessing science and technology in the region in specific areas, including the following, which Council, in its deliberations, noted were only examples of a wide array of possible initiatives:

- Available cost effective technology for promoting access to medical facilities particularly in rural areas should be utilised by Member States as appropriate, such as ultrasound technology and health-services that can be facilitated by mobile telephony;
- In the area of education, innovative initiatives for promoting access to education material, such as the [One Laptop per Child] project, currently in use worldwide, including in Rwanda, should be championed by COMESA Member States;
- In the life sciences, COMESA should utilise available information generated through the decoding and annotation of various genomes, to apply it in various areas such as developing crops that are adapted to the geographical conditions of the region;
- Noting that the available stock of technological knowledge increases exponentially, doubling every 12 months, and to take advantage of rapidly reducing costs of technological products, COMESA needs to develop mechanisms for harnessing relevant available technological knowledge worldwide; but for this to be possible, mechanisms should be put in place for developing the technical capacity to know and absorb the available knowledge worldwide in order to be able to apply it as appropriate in dealing with challenges that face the region in key priority areas such as agriculture, infrastructure, information and communications, public health, clean energy and water, environmental protection, and trade and economics. In particular, there is need to

mobilise and organise the region's scientists and engineers and encourage incremental innovation by individuals and SMEs;

- In the area of telecommunications, the various undersea and land cable networks for connecting up Africa and connecting Africa to the rest of the world should be utilised by Member States and stakeholders including the private sector, bearing in mind that Africa has significantly contributed financially to installing them; and
- COMESA should utilise the wireless broad band access that is going to be delivered in the tropics around the world by the set of 16 satellites being launched.

The Council welcomed the presentation and commended the Secretariat for arranging the presentation from such a brilliant son of Africa. The Council extensively deliberated the presentation and adopted Decisions. In terms of the way forward regarding the institutional framework, the Council noted the need for inter-ministerial coordination in order to avoid uncoordinated approaches, and in this regard, the vital importance of overarching executive leadership at a high level of Government. At the level of the Secretariat, if the Office of Advisor on Science and Technology is set up, it should be structured in a manner that ensures mainstreaming of science and technology in all the other programs and that avoids a silo approach, and it should have the primary objective of assisting Member States in their science and technology programs.

Decisions

The Council adopted the recommendations of the presentation and underscored the importance of mainstreaming science and technology in all COMESA programs and of adopting a cost effective approach that does not financially overburden the Member States and the Secretariat.

Furthermore, the Council urged Member States to:

- Promote the commercialisation of research and development, and put in place initiatives for improvement and standardisation of traditional products, innovating them into products that can be commercialised;
- Consider using biotechnology in the cropping sector in order to increase the outputs in the region, working with partners such as ECA and NEPAD, and taking into account the enormous biodiversity in the region;
- Dedicate at least 1% of the Gross Domestic Product to research and development, in line with the target set within the framework of the African Union;
- Consider adopting initiatives for promoting and utilising nano technology and science, given its application in various key areas such as medical treatment resulting from much higher levels of precision;
- Put in place concrete mechanisms for leveraging science and technology to address the key priorities in the region;
- Establish data bases for identifying individuals with the right profiles that can assist the implementation of science and technology initiatives in COMESA;
- Harmonise and coordinate their policy frameworks on science and technology at the COMESA level; and
- Elaborate and adopt master plans and blue prints for leveraging technological knowledge, for harnessing science and technology, and for mobilising the required resources.

Council decided also that:

- Member States should consider establishing Science and Technology Committees and Advisory Office at the highest level of Government;
- The Secretariat should establish an Office of Advisor on Science and Technology; and

- An Annual Innovation Award should be established to recognise outstanding accomplishment.

Broad Band Wireless Interactive System

Regarding a Broad Band Wireless Interactive System, the Council received a PowerPoint presentation, which was presented to the Fourth Meeting of the Ministers of Infrastructure at its meeting, 29–30 July 2010. The Report of that Meeting, reference CS/ID/MIN/IV/2, in paragraphs 384 to 388, provides the deliberations by the Infrastructure Ministers and a recommendation on this matter.

Recommendation

The Council noted the deliberations of the Infrastructure Ministers on this matter and endorsed the recommendation that pilot projects be developed to deploy the COBIS system in selected COMESA Member States following which when successfully implemented can be expanded for region-wide deployment.

Directives of the COMESA Heads of State and Government on Harnessing Science and Technology

The Heads of State and Government as well deliberated this matter of harnessing science and technology and listened to the video presentation of Professor Calestous Juma. They directed as follows, endorsing the Decisions of the Ministers:

Member States should:

- Where possible, pool resources and combine efforts to establish common science and technology parks;
- Promote the commercialisation of research and development, and put in place initiatives for improvement and

standardisation of traditional products, innovating them into products that can be commercialised;

- Consider using biotechnology in the cropping sector in order to increase the outputs in the region, working with partners such as ECA and NEPAD, and taking into account the enormous biodiversity in the region;
- Dedicate at least 1% of the Gross Domestic Product to research and development, in line with the target set within the framework of the African Union;
- Consider adopting initiatives for promoting and utilising nanotechnology and science, given its application in various key areas such as medical treatment resulting from much higher levels of precision;
- Develop a common curriculum in ICT that enables COMESA citizens to be exposed to ICT at an early age;
- Create a central fund that will concentrate on availing financial resources towards funding programs for ICT training and skills development;
- Establish data bases for identifying individuals with the right profiles that can assist the implementation of science and technology initiatives in COMESA;
- Harmonise and coordinate their policy frameworks on science and technology at the COMESA level; and
- Elaborate and adopt master plans and blue prints for leveraging technological knowledge, for harnessing science and technology, and for mobilising the required resources.

In addition:

- Member States should consider establishing Science and Technology Committees and Advisory Offices at the highest level of Government;
- The Secretariat should establish an Office of Advisor on Science and Technology;

- An Annual Innovation Award should be established to recognise outstanding accomplishment; and
- Member States should adopt a policy for harnessing science and technology.

Furthermore, the Heads of State and Government endorsed the COMESA Policy on Intellectual Property Rights and Cultural Industries as adopted by the Ministers.

NOTES

Introduction

1. B. wa Mutharika, "Feeding Africa through New Technologies: Let Us Act Now" (Acceptance Speech on Election of the Chairman of the Assembly of the African Union, Addis Ababa, January 31, 2010). As wa Mutharika explained: "I firmly believe that if we could agree that food security at the Africa level is a priority, then other priorities such as climate change, ICT, transport and infrastructure development would also become a necessity to enhance flow of information, movement of people, goods and services including the production and supply of agricultural inputs within and among nations, regions and the continent at large. I therefore propose that we consider investing in the construction of infrastructure to support food security. We need to build food storage facilities, new roads, railways, airlines, shipping industries as well as develop inter-state networks to ensure that we can move food surplus to deficit areas more efficiently and more cheaply."

2. C. P. Reij and E. M. Smaling, "Analyzing Successes in Agriculture and Land Management in Sub-Saharan Africa: Is Macro-Level Gloom Obscuring Micro-Level Change?" *Land Use Policy* 25, no. 3 (2008): 410–20.

3. Discussions on the role of innovation in development often ignore the role of engineering in development. For more details, see C. Juma, "Redesigning African Economies: The Role of Engineering in International Development" (Hinton Lecture, Royal Academy of Engineering, London, 2006); P. Guthrie, C. Juma, and H. Sillem, eds.,

Engineering Change: Towards a Sustainable Future in the Developing World. London: Royal Academy of Engineering, 2008.

4. H. T. Vesala and K. M. Vesala, "Entrepreneurs and Producers: Identities of Finnish Farmers in 2001 and 2006," *Journal of Rural Studies* 26, no. 1 (2010): 21–30.

5. The African Union (AU) recognizes the following RECs as the continent's economic integration building blocks: Community of Sahel Sahara States (CEN-SAD); Arab Maghreb Union (AMU); Economic Community of Central African States (ECCAS); Common Market of Eastern and Southern Africa (COMESA); Southern African Development Community (SADC); Intergovernmental Authority for Development (IGAD); Economic Community of West African States (ECOWAS); the East African Community (EAC). For descriptions, see Appendix I.

6. B. Gavin, "The Euro–Mediterranean Partnership: An Experiment in North–South–South Integration," *Intereconomics* 40, no. 6 (2005): 353–60.

7. A. Aghrout, "The Euro-Maghreb Economic Partnership: Trade and Investment Issues," *Journal of Contemporary European Studies* 17, no. 3 (2009): 353–67.

8. A.-N. Cherigui et al., "Solar Hydrogen Energy: The European-Maghreb Connection—A New Way of Excellence for Sustainable Development," *International Journal of Hydrogen Energy* 34, no. 11 (2009): 4934–40.

9. K. Kausch, "The End of the 'Euro-Mediterranean Vision,'" *International Affairs* 85, no. 5 (2009): 963–75.

10. C. Juma and I. Serageldin, *Freedom to Innovate: Biotechnology in Africa's Development*. Addis Ababa: African Union and New Partnership for Africa's Development, 2007.

11. C. Juma and I. Serageldin, *Freedom to Innovate: Biotechnology in Africa's Development*. Addis Ababa: African Union and New Partnership for Africa's Development, 2007, 116–17.

12. Royal Society of London. *Reaping the Benefits: Science and the Sustainable Intensification of Global Agriculture*. London: Royal Society of London 2009.

13. E. Kraemer-Mbula and W. Wamae, eds., *Innovation and the Development Agenda*. Paris: Organisation for Economic Co-operation and Development, 2010.

Chapter 1

1. U. Lele, "Structural Adjustment, Agricultural Development and the Poor: Some Lessons from the Malawian Experience," *World Development* 18, no. 9 (1990): 1207–19.

2. D. E. Sahn and J. Arulpragasam. "The Stagnation of Smallholder Agriculture in Malawi: A Decade of Structural Adjustment," *Food Policy* 16, no. 3 (1991): 219–34.

3. V. Quinn, "A History of the Politics of Food and Nutrition in Malawi: The Context of Food and Nutritional Surveillance," *Food Policy* 19, no. 3 (1994): 255–71.

4. G. Denning et al., "Input Subsidies to Improve Smallholder Maize Productivity in Malawi: Toward an African Green Revolution," *PLoS Biology* 7, no. 1 (2009): e1000023.

5. B. wa Mutharika, "Capitalizing on Opportunity" (paper presented at the World Economic Forum on Africa, Cape Town, South Africa, June 4, 2008).

6. B. Chinsinga, *Reclaiming Policy Space: Lessons from Malawi's Fertilizer Subsidy Programme*. London: Future Agricultures, 2007.

7. F. Ellis, *Fertiliser Subsidies and Social Cash Transfers: Complementary or Competing Instruments for Reducing Vulnerability to Hunger?* Johannesburg, South Africa: Regional Hunger and Vulnerability Programme, 2009.

8. H. Ndilowe, personal communication, Embassy of Malawi, Washington, DC, 2009.

9. G. Denning et al. "Input Subsidies to Improve Smallholder Maize Productivity in Malawi: Toward an African Green Revolution," *PLoS Biology* 7, no. 1 (2009): e1000023.

10. D. C. Chibonga, personal communication, National Smallholder Farmers' Association, Malawi, 2009.

11. D. C. Chibonga, personal communication, National Smallholder Farmers' Association, Malawi, 2009.

12. C. P. Timmer, "Agriculture and Economic Development," *Handbook of Agricultural Economics* 2 (2002): 1487–1546.

13. P. Pingali, "Agricultural Growth and Economic Development: A View through the Globalization Lens," *Agricultural Economics* 37, no. 1 (2007): 1–12.

14. D. Byerlee, A. de Janvy, and E. Sadoulet, "Agriculture for Development: Toward a New Paradigm," *Annual Review of Resource Economics* 1, no.1 (2009): 15–35.

15. P. Pingali, "Agriculture Renaissance: Making 'Agriculture for Development' Work in the 21st Century," *Handbook of Agricultural Economics* 4 (2010): 3867–94.

16. M. Tiffen, "Transition in Sub-Saharan Africa: Agriculture, Urbanization and Income Growth," *World Development* 31, no. 8 (2003): 1343–66.

17. P. Gollin, "Agricultural Productivity and Economic Growth," *Handbook of Agricultural Economics* 4 (2010): 3825–66.

18. For a comprehensive review of food-related debates, see R. Paarlberg, *Food Politics: What Everyone Needs to Know*. New York: Oxford University Press, 2010.

19. R. Evenson and D. Gollin, "Assessing the Impact of the Green Revolution, 1960–2000," *Science* 300, no. 5620 (2003): 758–62.

20. R. Evenson and D. Gollin, "Assessing the Impact of the Green Revolution, 1960–2000," *Science* 300, no. 5620 (2003): 758–62.

21. This is clearly articulated in InterAcademy Council, *Realizing the Promise and Potential of African Agriculture*. Amsterdam: InterAcademy Council, 2004.

22. P. Hazell, "An Assessment of the Impact of Agricultural Research in South Asia since the Green Revolution," *Handbook of Agricultural Economics* 4 (2010): 3469–3530.

23. World Bank, *World Development Report, 2008*. Washington, DC: World Bank, 2008, 27.

24. World Bank, *World Development Report, 2008*. Washington, DC: World Bank, 2008, 29.

25. World Bank, *World Development Report, 2008*. Washington, DC: World Bank, 2008, 40–41.

26. World Bank, *World Development Report, 2008*. Washington, DC: World Bank, 2008, 53.

27. World Bank, *World Development Report, 2008*. Washington, DC: World Bank, 2008, 27.

28. Organization for Economic Cooperation and Development, "Agriculture, Food Security and Rural Development for Growth and Poverty Reduction: China's Agricultural Transformation—Lessons for Africa and its Development Partners." Summary of discussions by the China–DAC Study Group, Bamako, Mali, April 27–28, 2010.

29. United Nations Economic Commission for Africa and African Union, *Economic Report on Africa 2009: Developing African Agriculture through Regional Value Chains*. Addis Ababa: United Nations Economic Commission for Africa and African Union, 2009.

30. UNEP-UNCTAD Task Force on Capacity Building on Trade, Environment and Development, *Organic Agriculture and Food Security in Africa*. Geneva: United Nations Conference on Trade and Development, 2008.

31. United Nations Economic Commission for Africa and African Union, *Economic Report on Africa 2009: Developing African Agriculture through Regional Value Chains*. Addis Ababa: United Nations Economic Commission for Africa and African Union, 2009.

32. E. Fabusoro et al. "Forms and Determinants of Rural Livelihood Diversification in Ogun State," *Journal of Sustainable Agriculture* 34, no. 4 (2010): 417–38.

33. United Nations Economic Commission for Africa and African Union, *Economic Report on Africa 2009: Developing African Agriculture through Regional Value Chains*. Addis Ababa: United Nations Economic Commission for Africa and African Union, 2009.

34. R. Paarlberg, *Starved for Science: How Biotechnology Is Being Kept Out of Africa*. Cambridge, MA: Harvard University Press, 2008.

35. United Nations Economic Commission for Africa and African Union, *Economic Report on Africa 2009: Developing African Agriculture through Regional Value Chains*. Addis Ababa: United Nations Economic Commission for Africa and African Union, 2009.

36. R. Roxburgh et al., *Lions on the Move: The Progress and Potential of African Economies*. Washington, DC: McKinsey Global Institute, 2010.

37. R. Roxburgh et al., *Lions on the Move: The Progress and Potential of African Economies*. Washington, DC: McKinsey Global Institute, 2010.

38. R. Roxburgh et al., *Lions on the Move: The Progress and Potential of African Economies*. Washington, DC: McKinsey Global Institute, 2010.

39. D. Deininger et al., *Rural Land Certification in Ethiopia: Process, Initial Impact, and Implications for Other African Countries*. Washington, DC: World Bank, 2007.

40. H. Binswanger-Mkhize and A. McCalla, "The Changing Context and Prospects for Agricultural and Rural Development in Africa," *Handbook of Agricultural Economics* 4 (2010): 3582–83.

41. P. D. Williams and J. Haacke, "Security Culture, Transnational Challenges and the Economic Community of West African States," *Journal of Contemporary African Studies* 26, no. 2 (2008): 119–23.

42. OECD, *African Economic Outlook 2010: Public Resource Mobilization and Aid*. Paris: Organisation for Economic Cooperation and Development, 2010.

43. EAC, *Agriculture and Rural Development Strategy for the East African Community*. Arusha, Tanzania: East African Community, 2006.

44. EAC, *Agriculture and Rural Development Strategy for the East African Community*. Arusha, Tanzania: East African Community, 2006.

45. T. Persson and G. Tabellini, "The Growth Effect of Democracy: Is It Heterogeneous and How Can It Be Estimated?" In *Institutions and Economic Performance*, ed. E. Helpman. Cambridge, MA: Harvard University Press, 2008.

46. L. E. Fulginiti, "What Comes First, Agricultural Growth or Democracy?" *Agricultural Economics* 41, no. 1 (2010): 15–24.

47. R. Nelson, "What Enables Rapid Economic Progress: What Are the Needed Institutions?" *Research Policy* 37, no.1 (2008): 1.

48. D. Restuccia, D. T. Yang, and X. Zhu, "Agriculture and Aggregate Productivity," *Journal of Monetary Economics* 55, no. 2 (2008): 234–50.

Chapter 2

1. J. Schwerin and C. Werker, "Learning Innovation Policy Based on Historical Experience," *Structural Change and Economic Dynamics* 14, no. 4 (2004): 385–404.

2. C. Freeman and F. Louçã, *As Time Goes By: From the Industrial Revolution to the Information Revolution*. Oxford: Oxford University Press, 2001; J. Fegerberg et al., eds., *The Oxford Handbook of Innovation*. Oxford: Oxford University Press, 2005.

3. T. Ridley, Y.-C. Lee, and C. Juma, "Infrastructure, Innovation and Development," *International Journal of Technology and Globalisation* 2, no. 3 (2006): 268–78; D. Rouach and D. Saperstein, "Alstom Technology Transfer Experience: The Case of the Korean Train Express (KTX)," *International Journal of Technology Transfer and Commercialisation* 3, no. 3 (2004): 308–23.

4. B. Oyelaran-Oyeyinka and K. Lal, "Learning New Technologies by Small and Medium Enterprises in Developing Countries," *Technovation* 26, no. 2 (2006): 220–31.

5. C. Juma, "Redesigning African Economies: The Role of Engineering in International Development" (Hinton Lecture, Royal Academy of Engineering, London, 2006).

6. D. King, "Governing Technology and Growth." In *Going for Growth: Science, Technology and Innovation in Africa*, ed. C. Juma. London: Smith Institute, 2005, 112–24.

7. C. Juma, "Reinventing African Economies: Technological Innovation and the Sustainability Transition" (paper presented at the 6th John Pesek Colloquium on Sustainable Agriculture, Iowa State University, Ames, Iowa, USA, 2006).

8. A. D. Alene, "Productivity Growth and the Effects of R&D in African Agriculture," *Agricultural Economics* 41, nos. 3–4 (2010): 223–38.

9. W. A. Masters, "Paying for Prosperity: How and Why to Invest in Agricultural R&D for Development in Africa," *Journal of International Affairs* 58, no. 2 (2005): 35–64.

10. H. Jikun and S. Rozelle, "Technological Change: Rediscovering the Engine of Productivity Growth in China's Rural Economy," *Journal of Development Economics* 49, no. 2 (1996): 337–69.

11. S. Fan, L. Zhang, and X. Zhang, "Reforms, Investment, and Poverty in Rural China," *Economic Development and Cultural Change* 5, no. 2 (2004): 395–422.

12. G. Conway and J. Waage, *Science and Innovation for Development.* London: UK Collaborative on Development Sciences, 2010, 37.

13. J. C. Acker and I. M. Mbiti, *Mobile Phones and Economic Development in Africa.* Washington, DC: Center for Global Development, 2010.

14. G. Conway and J. Waage, *Science and Innovation for Development.* London: UK Collaborative on Development Sciences, 2010, 37.

15. G. Conway and J. Waage, *Science and Innovation for Development.* London: UK Collaborative on Development Sciences, 2010, 37.

16. W. Jack and T. Suri, *Mobile Money: The Economics of M-PESA.* Cambridge, MA: Sloan School, Massachusetts Institute of Technology, 2009.

17. I. Mas, "The Economics of Branchless Banking," *Journal of Monetary Economics* 4, no. 2 (2009): 57–76.

18. M. L. Rilwani and I. A. Ikhuoria, "Precision Farming with Geoinformatics: A New Paradigm for Agricultural Production in a Developing Country," *Transactions in GIS* 10, no. 2 (2006): 177–97.

19. R. Chwala, personal communication, Survey Settlements and Land Records Department, State of Karnataka, 2009.

20. I. Potrykus, "Nutritionally Enhanced Rice to Combat Malnutrition Disorders of the Poor," *Nutrition Reviews* 61, Supplement 1 (2003): 101–4.

21. C. James, *Global Status of Commercialized Biotech/GM Crops: 2009.* ISAAA Brief No. 41. Ithaca, NY: International Service for the Acquisition of Agri-biotech Applications.

22. C. James, *Global Status of Commercialized Biotech/GM Crops: 2009.* ISAAA Brief No. 41. Ithaca, NY: International Service for the Acquisition of Agri-biotech Applications.
23. C. James, *Global Status of Commercialized Biotech/GM Crops: 2009.* ISAAA Brief No. 41. Ithaca, NY: International Service for the Acquisition of Agri-biotech Applications.
24. C. Pray, "Public-Private Sector Linkages in Research and Development: Biotechnology and the Seed Industry in Brazil, China and India," *American Journal of Agricultural Economics* 83, no. 3 (2001): 742–47.
25. R. Kaplinsky et al., "Below the Radar: What Does Innovation in Emerging Economies Have to Offer Other Low-income Countries?" *International Journal of Technology Management and Sustainable Development* 8, no. 3 (2009): 177–97. Indian entrepreneurs have figured ways of doing more with less based on the principles of affordability and sustainability: C. P. Prahalad and R. A. Mashelkar, "Innovation's Holy Grain," *Harvard Business Review*, July–August (2010): 1–10.
26. J. D. Glover and J. P. Reganold, "Perenial Grains: Food Security for the Future," *Issues in Science and Technology* 26, no. 2 (2010): 41–47.
27. C. James, *Global Status of Commercialized Biotech/GM Crops: 2009.* ISAAA Brief No. 41. Ithaca, NY: International Service for the Acquisition of Agri-biotech Applications.
28. A. M. Showalter et al., "A Primer for Using Transgenic Insecticidal Cotton in Developing Countries," *Journal of Insect Science* 9 (2009): 1–39.
29. D. Zilberman, H. Ameden, and M. Qaim, "The Impact of Agricultural Biotechnology on Yields, Risks, and Biodiversity in Low-Income Countries," *Journal of Development Studies* 43, no. 1 (2007): 63–78.
30. C. E. Pray et al. "Five Years of Bt Cotton in China—The Benefits Continue," *Plant Journal* 31, no. 4 (2000): 423–30.
31. R. Falkner, "Regulating Biotech Trade: The Cartagena Protocol on Biosafety," *International Affairs* 76, no. 2 (2000): 299–313.
32. L. R. Ghisleri et al., "Risk Analysis and GM Foods: Scientific Risk Assessment," *European Food and Feed Law Review* 4, no. 4 (2009): 235–50.
33. T. Bernauer, *Genes, Trade, and Regulation: The Seeds of Conflict in Food Biotechnology.* Princeton, NJ: Princeton University Press, 2003. Most

of the studies on the risks of agricultural biotechnology tend to focus on unintended negative impacts. But evidence of unintended benefits is emerging. See, for example, W. D. Hutchison et al., "Areawide Suppression of European Corn Borer with Bt Maize Reaps Savings to Non-Bt Maize Growers," *Science* 330, no. 6001 (2010): 222–25.

34. C. E. Pray et al., "Costs and Enforcement of Biosafety Regulations in India and China," *International Journal of Technology and Globalization* 2, nos. 1–2 (2006): 137–57; C. E. Pray, P. Bengali, and B. Ramaswami, "The Cost of Regulation: The India Experience," *Quarterly Journal of International Agriculture* 44, no. 3 (2005): 267–89.

35. R. Paarlberg, *Starved for Science: How Biotechnology Is Being Kept Out of Africa.* Cambridge, MA: Harvard University Press, 2008, 2.

36. E. van Kleef et al., "Food Risk Management Quality: Consumer Evaluations of Past and Emerging Food Safety Incidents," *Health, Risk and Society* 11, no. 2 (2009): 137–63.

37. L. J. Frewer et al., "What Determines Trust in Information about Food-related Risks? Underlying Psychological Constructs," *Risk Analysis* 16, no. 4 (1996): 473–85.

38. P. Jackson, "Food Stories: Consumption in the Age of Anxiety," *Cultural Geographies* 17, no. 2 (2010): 147–65.

39. S. Lieberman and T. Gray, "The World Trade Organization's Report on the EU's Moratorium on Biotech Products: The Wisdom of the US Challenge to the EU in the WTO," *Global Environmental Politics* 8, no. 1 (2008): 33–52.

40. N. Zerbe, "Feeding the Famine? American Food Aid and the GMO Debate in Southern Africa," *Food Policy* 29, no. 6 (2004): 593–608.

41. I. Cheyne, "Life after the Biotech Products Dispute," *Environmental Law Review* 10, no. 1 (2008): 52–64.

42. E. J. Morris, "The Cartagena Protocol: Implications for Regional Trade and Technology Development in Africa," *Development Policy Review* 26, no. 1 (2008): 29–57.

43. D. Wield, J. Chataway, and M. Bolo, "Issues in the Political Economy of Agricultural Biotechnology," *Journal of Agrarian Change* 10, no. 3 (2010): 342–66.

44. J. O. Adeoti and A. A. Adekunle, "Awareness of and Attitudes towards Biotechnology and GMOs in Southwest Nigeria: A Survey of People with Access to Information," *International Journal of Biotechnology* 9, no. 2 (2007): 209–30.

45. E. J. Morris, "The Cartagena Protocol: Implications for Regional Trade and Technology Development in Africa," *Development Policy Review* 26, no. 1 (2008): 29–57.

46. C. Bett, J. O. Ouma, and H. Groote, "Perspectives on Gatekeepers in the Kenya Food Industry toward Genetically Modified Food," *Food Policy* 35, no. 4 (2010): 332–40.

47. D. L. Kleinman, A. J. Kinchy, and R. Autry, "Local Variations or Global Convergence in Agricultural Biotechnology Policy? A Comparative Analysis," *Science and Public Policy* 36, no. 5 (2009): 361–71.

48. J. Keeley, "Balancing Technological Innovation and Environmental Regulation: An Analysis of Chinese Agricultural Biotechnology Governance," *Environmental Politics* 15, no. 2 (2000): 293–309.

49. National Research Council. *Impact of Genetically Engineered Crops on Farm Sustainability in the United States.* Washington, DC: National Academies Press, 2010.

50. G. Conway and J. Waage, *Science and Innovation for Development.* London: UK Collaborative on Development Sciences, 2010, 54.

51. T. Fleischer and A. Grunwald, "Making Nanotechnology Developments Sustainable: A Role for Technology Assessment?" *Journal of Cleaner Production* 16, nos. 8–9 (2008): 889–98.

52. G. Conway and J. Waage, *Science and Innovation for Development.* London: UK Collaborative on Development Sciences, 2010, 53.

53. S. Louis and D. Alcorta, personal communication, Dais Analytic Corporation, Odessa, Florida, USA (2010).

54. J. Perkins, *Geopolitics and the Green Revolution: Wheat, Genes, and the Cold War.* New York: Oxford University Press, 1997.

55. B. Bell Jr. and C. Juma, "Technology Prospecting: Lessons from the Early History of the Chile Foundation." *International Journal of Technology and Globalisation* 3, nos. 2–3 (2007): 296–314.

56. K. Kastenhofer, "Do We Need a Specific Kind of Technoscience Assessment? Taking the Convergence of Science and Technology Seriously," *Poiesis Prax* 7, nos. 1–2 (2010): 37–54.

57. W. Russell et al., "Technology Assessment in Social Context: The Case for a New Framework for Assessing and Shaping Technological Development," *Impact Assessment and Project Appraisal* 28, no. 2 (2010): 109–16.

58. L. Pellizoni, "Uncertainty and Participatory Democracy," *Environmental Values* 12, no. 2 (2003): 195–224.

59. A. Hall, "Embedding Research in Society: Development Assistance Options for Agricultural Innovation in a Global Knowledge

Economy," *International Journal of Technology Management and Sustainable Development* 8, no. 3 (2008): 221–35.

Chapter 3

1. G. E. Glasson et al., "Sustainability Science Education in Africa: Negotiating Indigenous Ways of Living with Nature in the Third Space," *International Journal of Science Education* 32, no. 1 (2010): 125–41.

2. S. W. Omamo and J. K. Lynam, "Agricultural Science and Technology Policy in Africa," *Research Policy* 32, no. 9 (2003): 1681–94.

3. J. Fagerberg, "Introduction: A Guide to the Literature." In *The Oxford Handbook of Innovation*, ed. J. Fagerberg, D. Mowery, and R. Nelson. Oxford: Oxford University Press, 2005, 1–26.

4. J. Sumberg, "Systems of Innovation Theory and the Changing Architecture of Agricultural Research in Africa," *Food Policy* 30, no. 1 (2005): 21–41.

5. A. Hall, W. Janssen, E. Pehu, and R. Rajalahti, *Enhancing Agricultural Innovation: How to Go Beyond Strengthening Research Systems*. Washington, DC: World Bank, 2006.

6. A. Hall, W. Janssen, E. Pehu, and R. Rajalahti, *Enhancing Agricultural Innovation: How to Go Beyond Strengthening Research Systems*. Washington, DC: World Bank, 2006.

7. D. J. Spielman et al., "Developing the Art and Science of Innovation Systems Enquiry: Alternative Tools and Methods, and Applications to Sub-Saharan African Agriculture," *Technology in Society* 31, no. 4 (2009): 399–405.

8. J. O. Adeoti and O. Olubamiwa, "Toward an Innovation System in the Traditional Sector: The Case of the Nigerian Cocoa Industry," *Science and Public Policy* 36, no. 1 (2009): 15–31.

9. J. O. Adeoti, S. O. Odekunle, and F. M. Adeyinka, *Tackling Innovation Deficit: An Analysis of University-Firm Interaction in Nigeria*. Ibadan, Nigeria: Evergreen, 2010.

10. For more details, see J. Adeoti and O. Olubamiwa, "Toward an Innovation System in the Traditional Sector: The Case of the Nigerian Cocoa Industry," *Science and Public Policy* 36, no. 1 (2009): 15–31.

11. M. Gagné et al., "Technology Cluster Evaluation and Growth Factors: Literature Review," *Research Evaluation* 19, no. 2 (2010): 82–90.

12. M. S. Gertler and T. Vinodrai, "Life Sciences and Regional Innovation: One Path or Many?" *European Planning Studies* 17, no. 2 (2009): 235–61.

13. M. E. Porter, "Clusters and the New Economics of Competition," *Harvard Business Review* 76, no. 6 (1998): 77–90.

14. Z. Yingming, *Analysis of Industrial Clusters in China.* Boca Raton, Florida, USA: CRC Press, 2010.

15. This case study draws heavily from G. Wu, T. Tu and S. Gu, "Innovation System and Transformation of the Agricultural Sector in China, with the Case of Shouguang City" (paper presented to the Globelics Conference on Innovation and Development, Rio de Janeiro, November 3–6, 2003); S. Gu, "The Emergence and Development of the Vegetable Sector in China," *Industry and Innovation* 26, nos. 4–5 (2009): 499–524.

16. D. Dalohoun, A. Hall, and P. Mele, "Entrepreneurship as Driver of a 'Self-Organizing System of Innovation': The Case of NERICA in Benin," *International Journal of Technology Management and Sustainable Development* 8, no. 2 (2009): 87–101.

17. M. Dermastia, "Slovenia." In *Business Clusters: Promoting Enterprise in Central and Eastern Europe*, ed. OECD. Paris: Organization for Economic Cooperation and Development, 2005.

18. P. Maskell, "Towards a Knowledge-Based Theory of the Geographical Cluster," *Industrial and Corporate Change* 10, no. 4 (2001): 921.

19. S. Breschi and F. Malerba. "The Geography of Innovation and Economic Clustering: Some Introductory Notes," *Industrial and Corporate Change* 10, no. 4 (2001): 817.

20. B. Orlove et al., "Indigenous Climate Change Knowledge in Southern Uganda," *Climatic Change* 100, no. 2 (2010): 243–65.

21. J. F. Martin et al., "Traditional Ecological Knowledge (TEK): Ideas, Inspiration, and Design for Ecological Engineering," *Ecological Engineering* 36, no. 7 (2010): 839–49.

22. G. Glasson et al., "Sustainability Science Education in Africa: Negotiating Indigenous Ways of Living with Nature in the Third Space," *International Journal of Science Education* 32, no. 1 (2010): 125–41.

23. E. Ostrom, *Understanding Institutional Diversity.* Princeton, NJ: Princeton University Press, 2005.

24. Z. Xiwei and Y. Xiangdong, "Science and Technology Policy Reform and Its Impact on China's National Innovation System," *Technology in Society* 29, no. 3 (2007): 317.

25. Z. Xiwei and Y. Xiangdong, "Science and Technology Policy Reform and Its Impact on China's National Innovation System," *Technology in Society* 29, no. 3 (2007): 321.

26. X. Liu and T. Zhi, "China Is Catching Up in Science and Innovation: The Experience of the Chinese Academy of Sciences," *Science and Public Policy* 37, no. 5 (2010): 331–42.

27. Z. Xiwei and Y. Xiangdong, "Science and Technology Policy Reform and Its Impact on China's National Innovation System," *Technology in Society* 29, no. 3 (2007): 322.

28. M. A. Lopes and P. B. Arcuri, *The Brazilian Agricultural Research for Development (ARD) System* (paper presented at the International Workshop on Fast Growing Economies' Role in Global Agricultural Research for Development [ARD], Beijing, China, February 8–10, 2010).

Chapter 4

1. E. B. Barrios, "Infrastructure and Rural Development: Household Perceptions on Rural Development," *Progress in Planning* 70, no. 1 (2008): 1–44.

2. P. R. Agenor, "A Theory of Infrastructure-led Development," *Journal of Economic Dynamics and Control* 34, no. 5 (2010): 932–50.

3. R. G. Teruel and Y. Kuroda, "Public Infrastructure and Productivity Growth in Philippine Agriculture, 1974–2000," *Journal of Asian Economics* 16, no. 3 (2005): 555–76; P.-R. Agenor and B. Moreno-Dodson, *Public Infrastructure and Growth: New Channels and Policy Implications.* Washington, DC: World Bank, 2006.

4. S. Fan and X. Zhang, "Public Expenditure, Growth and Poverty Reduction in Rural Uganda," *Africa Development Review* 20, no. 3 (2008): 466–96.

5. S. Fan and C. Chan-Kang, *Road Development, Economic Growth, and Poverty Reduction in China.* Washington, DC: International Food Policy Research Institute, 2005.

6. A. Narayanamoorthy, "Economics of Drip Irrigation in Sugarcane Cultivation: Case Study of a Farmer from Tamil Nadu," *Indian Journal of Agricultural Economics* 60, no. 2 (2005): 235.

7. The rest of this section is based on K. N. Gratwick and A. Eberhard, "An Analysis of Independent Power Projects in Africa: Understanding Development and Investment Outcomes," *Development Policy Review* 26, no. 3 (2008): 309–38.

8. K. Annamalai and S. Rao, *ITC's e-Choupal and the Profitable Rural Transformation*. Washington, DC: World Resource Institute (WRI), Michigan Business School, and UNC Kenan Flagler Business School.

9. This section draws heavily from D. Rouach and D. Saperstein, "Alstom Technology Transfer Experience: The Case of the Korean Train Express (KTX)," *International Journal of Technology Transfer and Commercialisation* 3, no. 3 (2004): 308–23.

10. T. E. Mutambara, "Regional Transport Challenges with the South African Development Community and Their Implications for Economic Integration and Development,"*Journal of Contemporary African Studies* 27, no. 4 (2009): 501–25.

11. E. Calitz and J. Fourie, "Infrastructure in South Africa: Who Is to Finance and Who Is to Pay?" *Development Southern Africa* 27, no. 2 (2010): 177–91.

12. K. Putranto, D. Stewart, and G. Moore, "International Technology Transfer and Distribution of Technology Capabilities: The Case of Railway Development in Indonesia," *International Journal of Technology Transfer and Commercialisation* 3, no. 3 (2003): 43–53.

13. R. Shah and R. Batley, "Private-Sector Investment in Infrastructure: Rationale and Causality for Pro-poor Impacts," *Development Policy Review* 27, no. 4 (2009): 397–417.

Chapter 5

1. A. Pinkerton and K. Dodds, "Radio Geopolitics: Broadcasting, Listening and the Struggle for Acoustic Space,"*Progress in Human Geography* 33, no. 1 (2009): 10–27.

2. N. T. Assié-Lumumba, *Empowerment of Women in Higher Education in Africa: The Role and Mission of Research*. Paris: United Nations Educational, Scientific and Cultural Organization, 2006.

3. M. Slavik, "Changes and Trends in Secondary Agricultural Education in the Czech Republic," *International Journal of Educational Development* 24, no. 5 (2004): 539–45.

4. J. Ndjeunga et al., *Early Adoption of Modern Groundnut Varieties in West Africa*. Hyderabad, India: International Crops Research in Semi-Arid Tropics, 2008.

5. G. Feder, R. Murgai, and J. Quizon, "Sending Farmers Back to School: The Impact of Farmer Field Schools in Indonesia," *Review of Agricultural Economics* 26, no. 1 (2004): 45–62.

6. P. Woomer, M. Bokanga, and G. Odhiambo, "*Striga* Management and the African Farmer," *Outlook on Agriculture* 37, no. 4 (2008): 277–82.

7. G. McDowell, Land-Grant Universities and Extension into the 21st Century: Renegotiating or Abandoning a Social Contract. Ames: Iowa State University Press, 2001.

8. H. Etzkowitz, "The Evolution of the Entrepreneurial University," *International Journal of Technology and Globalisation* 1, no. 1 (2004): 64–77.

9. M. Almeida, "Innovation and Entrepreneurship in Brazilian Universities," *International Journal of Technology Management and Sustainable Development* 7, no. 1 (2008): 39–58.

10. W. J. Mitsch et al., "Tropical Wetlands for Climate Change Research, Water Quality Management and Conservation Education on a University Campus in Costa Rica," *Ecological Engineering* 34, no. 4 (2008): 276–88.

11. The details on EARTH University are derived from C. Juma, "Agricultural Innovation and Economic Growth in Africa: Renewing International Cooperation," *International Journal of Technology and Globalisation* 4, no. 3 (2008): 256–75.

12. M. Miller, M. J. Mariola, and D. O. Hansen, "EARTH to Farmers: Extension and the Adoption of Environmental Technologies in Humid Tropics of Costa Rica," *Ecological Engineering* 34, no. 4 (2008): 349–57.

13. National Research Council, *Transforming Agricultural Education for a Changing World*. Washington, DC: National Academies Press, 2009.

14. A. deGrassi, "Envisioning Futures of African Agriculture: Representation, Power, and Socially Constituted Time," *Progress in Development Studies* 7, no. 2 (2007): 79–98.

15. D. E. Leigh and A. M. Gill, "How Well Do Community Colleges Respond to the Occupational Training Needs of Local Communities? Evidence from California," *New Directions for Community Colleges* 2009, no. 146 (2009): 95–102.

16. M. Carnoy and T. Luschei, "Skill Acquisition in 'High Tech' Export Agriculture: A Case Study of Lifelong Earning in Peru's Asparagus Industry," *Journal of Education and Work* 21, no. 1 (2008): 1–23.

Chapter 6

1. M. Pragnell, "Agriculture, Business and Development," *International Journal of Technology and Globalisation* 2, nos. 3–4 (2006): 289–99.

2. J. Lerner, *Boulevard of Broken Dreams: Why Public Efforts to Boost Entrepreneurship and Venture Capital Have Failed—and What to Do About It*. Princeton, NJ: Princeton University Press, 2009.

3. A. Zacharakis, D. A. Shepherd, and J. A. Coombs, "The Development of Venture-Capital-Backed Internet Companies: An Ecosystem Perspective," *Journal of Business Venturing* 18, no. 2 (2003): 217–31.

4. G. Avnimelech, A. Rosiello, and M. Teubal, "Evolutionary Interpretation of Venture Capital Policy in Israel, Germany, UK and Scotland," *Science and Public Policy* 37, no. 2 (2010): 101–12.

5. Y. Huang, *Capitalism with Chinese Characteristics: Entrepreneurship and the State.* New York: Cambridge University Press, 2008.

6. Y. Huang, *Capitalism with Chinese Characteristics: Entrepreneurship and the State.* New York: Cambridge University Press, 2008.

7. Z. Zhang, "Rural Industrialization in China: From Backyard Furnaces to Township and Village Enterprises," *East Asia* 17 no. 3 (1999): 61–87.

8. S. Rozelle, J. Huang, and L. Zhang, "Emerging Markets, Evolving Institutions, and the New Opportunities for Growth in China's Rural Economy," *China Economic Review* 13, no. 4 (2002): 345–53.

9. H. Chen and S. Rozelle, "Leaders, Managers, and the Organization of Township and Village Enterprises in China," *Journal of Development Economics* 60, no. 2 (1999): 529–57.

10. N. Minot et al., "Seed Development Programs in Sub-Saharan Africa: A Review of Experiences" (paper prepared for the Rockefeller Foundation, Nairobi, 2007).

11. A. S. Langyituo et al., "Challenges of the Maize Seed Industry in Eastern and Southern Africa: A Compelling Case for Private-Public Intervention to Promote Growth," *Food Policy* 35, no. 4 (2010): 323–31.

12. C. E. Pray and L. Nagarajan, *Pearl Millet and Sorghum Improvement in India.* Washington, DC: International Food Policy Research Institute, 2009.

13. M. Blackie, "Output to Purpose Review: Seeds of Development Programme (SODP)" (paper commissioned by the UK Department for International Development, London, August 7, 2008).

14. J. Wilkinson, "The Food Processing Industry, Globalization and Development Economics," *Journal of Agricultural and Development Economics* 1, no. 2 (2004): 184–201.

15. A. Dannson et al., *Strengthening Farm-Agribusiness Linkages in Africa: Summary Results of Five Country Studies in Ghana, Nigeria, Kenya,*

Uganda and South Africa. Rome: Food and Agriculture Organization of the United Nations, 2004.

16. A. Dannson et al., *Strengthening Farm-Agribusiness Linkages in Africa: Summary Results of Five Country Studies in Ghana, Nigeria, Kenya, Uganda and South Africa*. Rome: Food and Agriculture Organization of the United Nations, 2004.

17. A. Dannson et al., *Strengthening Farm-Agribusiness Linkages in Africa: Summary Results of Five Country Studies in Ghana, Nigeria, Kenya, Uganda and South Africa*. Rome: Food and Agriculture Organization of the United Nations, 2004.

18. E. Zossou et al., "The Power of Video to Trigger Innovation: Rice Processing in Central Benin," *International Journal of Agricultural Sustainability* 7, no. 2 (2009): 119–29.

19. P. Van Mele, J. Wanvoeke, and E. Zossou, "Enhancing Rural Learning, Linkages, and Institutions: The Rice Videos in Africa," *Development in Practice* 20, no. 3 (2010): 414–421.

20. J. Kerlin, "A Comparative Analysis of the Global Emergence of Social Enterprise," *Voluntas: International Journal of Voluntary and Nonprofit Organizations* 21, no. 2 (2010): 162–79.

21. The rest of this section is based on S. Hanson, personal communication, Bungoma, Kenya, One Acre Fund.

22. A. Gupta, personal communication, Society for Research and Initiatives for Sustainable Technologies and Institutions, Ahmedabad, India, 2010.

Chapter 7

1. M. Farrell, "EU Policy Towards Other Regions: Policy Learning in the External Promotion of Regional Integration," *Journal of European Public Policy* 16, no. 8 (2009): 1165–84.

2. C. Juma and I. Serageldin, *Freedom to Innovate: Biotechnology in Africa's Development*. Addis Ababa: African Union and New Partnership for Africa's Development, 2007, 20.

3. C. Juma, C. Gitta, and A. DiSenso, "Forging New Technological Alliances: The Role of South-South Cooperation," *Cooperation South Journal* (2005): 59–71.

4. Commission for Africa, *Our Common Interest: Report of the Commission for Africa*. London: Commission for Africa, 2005.

5. Y. H. Kim, "The Optimal Path of Regional Economic Integration between Asymmetric Countries in the North East Asia," *Journal of Policy Modeling* 27, no. 6 (2005): 673–87.

6. C. Wagner, *The New Invisible College: Science for Development.* Washington, DC: Brookings Institution, 2008.

7. T. Paster, *The HACCP Food Safety Training Manual.* Hoboken, NJ: John Wiley & Sons, 2007.

8. G. Jones and S. Corbridge, "The Continuing Debate about Urban Bias: The Thesis, Its Critics, Its Influence and Its Implications for Poverty-Reduction Strategies," *Progress in Development Studies* 10, no. 1 (2010): 1–18.

9. L. M. Póvoa and M. S. Rapini, "Technology Transfer from Universities and Public Research Institutes to Firms in Brazil: What Is Transferred and How the Transfer Is Carried Out," *Science and Public Policy* 37, no. 2 (2010): 147–59.

10. S. Pitroda, personal communication, New Delhi: National Innovation Council (2010).

11. C. Juma and Y.-C. Lee, *Innovation: Applying Knowledge in Development.* London: Earthscan, 2005, 140–58.

12. A. Sindzingre, "Financing the Developmental State: Tax and Revenue Issues," *Development Policy Review* 25, no. 5 (2007): 615–32.

13. N. Bloom, R. Griffith, and J. Van Reenen, "Do R&D Tax Credits Work? Evidence from a Panel of Countries 1979–1997," *Journal of Public Economics* 85, no. 1 (2002): 1–31.

14. B. Aschhoff and W. Sofka, "Innovation on Demand—Can Public Procurement Drive Market Success of Innovation?" *Research Policy* 38, no. 8 (2009): 1235–47.

15. McKinsey & Company, *"And the Winner Is . . .": Capturing the Promise of Philanthropic Prizes.* McKinsey & Company, 2009. http://www.mckinsey.com/App_Media/Reports/SSO/And_the_winner_is.pdf.

16. W. A. Masters and B. Delbecq, *Accelerating Innovation with Prize Rewards.* Washington, DC: International Food Policy Research Institute, 2008.

17. T. Cason, W. Masters, and R. Sheremeta, "Entry into Winner-Take-All and Proportional-Prize Contests: An Experimental Study," *Journal of Public Economics* 94, nos. 9–10 (2010): 604–11.

18. Philip G. Pardey and P. Pingali, "Reassessing International Agricultural Research for Food and Agriculture"(report prepared for the Global Conference on Agricultural Research for Development

[GCARD], Montpellier, France, March 28–31, 2010); J. M. Alston, M. A. Andersen, J. S. James, and P. G. Pardey, *Persistence Pays: U.S. Agricultural Productivity Growth and the Benefits from Public R&D Spending*. New York: Springer, 2010.

19. R. E. Just, J. M. Alston, and D. Zilberman, eds., *Regulating Agricultural Biotechnology: Economics and Policy*. New York: Springer–Verlag, 2006.

20. W. B. Arthur, *The Nature of Technology: What It Is and How It Evolves*. New York: Free Press, 2009.

21. D. Archibugi and C. Pietrobelli, "The Globalisation of Technology and Its Implications for Developing Countries," *Technological Forecasting and Social Change* 70, no. 9 (2003): 861–83.

22. Y. Chen and T. Puttitanum, "Intellectual Property Rights and Innovation in Developing Countries," *Journal of Development Economics* 78, no. 2 (2005): 474–93.

23. C. E. Pray and N. Anwar, "Supplying Crop Biotechnology to the Poor: Opportunities and Constraints," *Journal of Development Studies* 43, no. 1 (2007): 192–217.

24. S. Redding and P. Schott, "Distance, Skill Deepening and Development: Will Peripheral Countries Ever Get Rich?" *Journal of Development Economics* 72, no. 2 (2003): 515–541.

25. Ç. Özden and M. Schiff, eds., *International Migration, Remittances and the Brain Drain*. Washington, DC: World Bank, 2005.

26. S. Carr, I. Kerr, and K. Thorn, "From Global Careers to Talent Flow: Reinterpreting 'Brain Drain,'" *Journal of World Business* 40, no. 4 (2005): 386–98.

27. O. Stark, "Rethinking the Brain Drain," *World Development* 32, no. 1 (2004): 15–22.

28. A. Saxenian, "The Silicon Valley-Hsinchu Connection: Technical Communities and Industrial Upgrading," *Industrial and Corporate Change* 10, no. 4 (2001): 893–920.

29. National Knowledge Commission, *Report to the Nation, 2006–2009*. New Delhi: National Knowledge Commission, Government of India.

30. M. MacGregor, Y. F. Adam, and S. A. Shire, "Diaspora and Development: Lessons from Somaliland," *International Journal of Technology and Globalisation* 4, no. 3 (2008): 238–55.

31. A. Leather et al., "Working Together to Rebuild Health Care in Post-Conflict Somaliland," *Lancet* 368, no. 9541 (2006): 1119–25.

32. M. Anas and S. Wickremasinghe, "Brain Drain of the Scientific Community of Developing Countries: The Case of Sri Lanka," *Science and Public Policy* 37, no. 5 (2010): 381.

33. N. Fedoroff, "Science Diplomacy in the 21st Century," *Cell* 136, no. 1 (2009): 9–11.

34. E. Chalecki, "Knowledge in Sheep's Clothing: How Science Informs American Diplomacy," *Diplomacy and Statecraft* 19, no. 1 (2008): 1–19.

35. A. H. Zewail, "Science in Diplomacy," *Cell* 141, no. 2 (2010): 204–7.

36. T. Shaw, A. Cooper, and G. Chin, "Emerging Powers and Africa: Implications for/from Global Governance," *Politikon* 36, no.1 (2009): 27–44.

37. A. de Freitas Barbosa, T. Narciso, and M. Biancalana, "Brazil in Africa: Emerging Power in the Continent?" *Politikon* 36, no.1 (2009): 59–86.

38. F. El-Baz, "Science Attachés in Embassies," *Science* 329, no. 5987 (July 2, 2010): 13.

39. B. Séguin, et al., "Scientific Diasporas as an Option for Brain Drain: Re-circulating Knowledge for Development," *International Journal of Biotechnology* 8, nos. 1–2 (2006): 78–90.

40. R. L. Tung, "Brain Circulation, Diaspora, and International Competitiveness," *European Management Journal* 26, no. 5 (2008): 298–304.

41. House of Commons Science and Technology Committee, *The Use of Science in UK International Development Policy*, Vol. 1. London: Stationery Office Limited, 2004.

42. T. Yakushiji, "The Potential of Science and Technology Diplomacy," *Asia-Pacific Review* 16, no.1 (2009): 1–7.

43. National Research Council, *The Fundamental Role of Science and Technology in International Development: An Imperative for the US Agency for International Development*. Washington, DC: National Academies Press, 2006.

44. D. D. Stine, *Science, Technology, and American Diplomacy: Background and Issues for Congress*. Washington, DC: Congressional Research Service, Congress of the United States.

45. K. King, "China's Cooperation in Education and Training with Kenya: A Different Model?" *International Journal of Educational Development* 30, no. 5 (2010): 488–96.

46. R. A. Austen and D. Headrick. "The Role of Technology in the African Past," *African Studies Review* 26, nos. 3–4 (1983): 163–84.

47. S. E. Cozzens, "Distributive Justice in Science and Technology Policy," *Science and Public Policy* 34, no. 2 (2007): 85–94.

48. J. Mugwagwa, "Collaboration in Biotechnology Governance: Why Should African Countries Worry about Those among Them that Are

Technologically Weak?" *International Journal of Technology Management and Sustainable Development* 8, no. 3 (2009): 265–79.

Chapter 8

1. P. K. Thornton et al., "Special Variation of Crop Yield Response to Climate Change in East Africa," *Global Environmental Change* 19, no. 1 (2009): 54–65.

2. World Bank, *World Development Report 2010: Development and Climate Change*. Washington, DC: World Bank, 2010, 5.

3. E. Bryan et al., "Adaptation to Climate Change in Ethiopia and South Africa: Options and Constraints," *Environmental Science and Policy* 12, no. 4 (2009): 413–26.

4. P. Laux et al., "Impact of Climate Change on Agricultural Productivity under Rainfed Conditions in Cameroon—a Method to Improve Attainable Crop Yields by Planting Date Adaptation," *Agricultural and Forest Meteorology* 150, no. 9 (2010): 1258–71.

5. W. N. Adger et al., "Assessment of Adaptation Practices, Options, Constraints and Capacity." In *Climate Change 2007: Impacts, Adaptation and Vulnerability. Contribution of Working Group II to the Fourth Assessment Report of the Intergovernmental Panel on Climate Change*, ed. M. L. Parry. Cambridge: Cambridge University Press, 2007, 869.

6. W. N. Adger et al. "Assessment of Adaptation Practices, Options, Constraints and Capacity." In *Climate Change 2007: Impacts, Adaptation and Vulnerability. Contribution of Working Group II to the Fourth Assessment Report of the Intergovernmental Panel on Climate Change*, ed. M. L. Parry. Cambridge: Cambridge University Press, 2007, 869.

7. M. Burke, D. Lobell, and L. Guarino, "Shifts in African Crop Climates by 2050, and the Implications for Crop Improvement and Genetic Resources Conservation," *Global Environmental Change* 19, no. 3 (2009): 317–25.

8. M. Burke, D. Lobell, and L. Guarino, "Shifts in African Crop Climates by 2050, and the Implications for Crop Improvement and Genetic Resources Conservation," *Global Environmental Change* 19, no. 3 (2009): 317–25.

9. P. G. Jones and P. K. Thornton, "Croppers to Livestock Keepers: Livelihood Transitions to 2050 in Africa due to Climate Change," *Environmental Science and Policy* 12, no. 4 (2009): 427–37.

10. S. N. Seo and R. Mendelsohn, "An Analysis of Crop Choice: Adapting to Climate Change in South American Farms," *Ecological Economics* 67, no. 1 (2008): 109–16.

11. N. Heller and E. Zavaleta, "Biodiversity Management in the Face of Climate Change: A Review of 22 Years of Recommendations," *Biological Conservation* 142, no. 1 (2009): 14–32.

12. M. Snoussi et al., "Impacts of Sea-level Rise on the Moroccan Coastal Zone: Quantifying Coastal Erosion and Flooding in the Tangier Bay," *Geomorphology* 107, nos. 1–2 (2009): 32–40.

13. M. Koetse and P. Rietveld, "The Impact of Climate Change and Weather on Transport: An Overview of Empirical Findings," *Transportation Research Part D: Transport and Environment* 14, no. 3 (2009): 205–21.

14. National Academy of Engineering, *Grand Challenges for Engineering*. Washington, DC: National Academies Press, 2008.

15. M. Keim, "Building Human Resilience: The Role of Public Health Preparedness and Response as an Adaptation to Climate Change," *American Journal of Preventive Medicine* 35, no. 5 (2008): 508–16.

16. K. Ebi and I. Burton, "Identifying Practical Adaptation Options: An Approach to Address Climate Change-related Health Risks," *Environmental Science and Policy* 11, no. 4 (2008): 359–69.

17. Y. Pappas and C. Seale, "The Opening Phase of Telemedicine Consultations: An Analysis of Interaction," *Social Science and Medicine* 68, no. 7 (2009): 1229–37.

18. T. Richmond, "The Current Status and Future Potential of Personalized Diagnostics: Streamlining a Customized Process," *Biotechnology Annual Review* 14 (2008): 411–22.

19. T. Deressa et al., "Determinants of Farmers' Choice of Adaptation Methods to Climate Change in the Nile Basin of Ethiopia," *Global Environmental Change* 19, no. 2 (2009): 248–55.

20. J. Youtie, and P. Shapira, "Building an Innovation Hub: A Case Study of the Transformation of University Roles in Regional Technological and Economic Development," *Research Policy* 37, no. 8 (2008): 1188–1204.

21. W. Hong. "Decline of the Center: The Decentralizing Process of Knowledge Transfer of Chinese Universities from 1985 to 2004," *Research Policy* 37, no. 4 (2008): 580–95.

22. S. Coulthard, "Adapting to Environmental Change in Artisanal Fisheries—Insights from a South Indian Lagoon," *Global Environmental Change* 18, no. 3 (2008): 479–89.

23. H. Osbahr et al., "Effective Livelihood Adaptation to Climate Change Disturbance: Scale Dimensions of Practice in Mozambique," *Geoforum* 39, no. 6 (2008): 1951–64.

24. J. Comenetz and C. Caviedes, "Climate Variability, Political Crises, and Historical Population Displacements in Ethiopia," *Global Environmental Change Part B: Environmental Hazards* 4, no. 4 (2002): 113–27.

25. C. S. Hendrix and S. M. Glaser, "Trends and Triggers: Climate, Climate Change and Civil Conflict in Sub-Saharan Africa," *Political Geography* 26, no. 6 (2007): 695–715.

26. M. Pelling and C. High, "Understanding Adaptation: What Can Social Capital Offer Assessments of Adaptive Capacity?" *Global Environmental Change Part A* 15, no. 4 (2005): 308–19.

27. H. de Coninck et al., "International Technology-oriented Agreements to Address Climate Change," *Energy Policy* 36, no. 1 (2008): 335–56.

28. S. N. Seo. "Is an Integrated Farm More Resilient against Climate Change? A Micro-econometric Analysis of Portfolio Diversification in African Agriculture," *Food Policy* 35, no. 1 (2010): 32–40.

29. N. von Tunzelmann, "Historical Coevolution of Governance and Technology in the Industrial Revolutions," *Structural Change and Economic Dynamics* 14, no. 4 (2003): 365–84.

30. M. Mintrom and P. Norman, "Policy Entrepreneurship and Policy Change," *Policy Studies Journal* 37, no. 4 (2009): 649–67.

31. Science and Technology Committee, House of Commons, *Scientific Advice, Risk and Evidence Based Policy Making*. London: Stationery Office, 2006.

32. J. Schwerin and C. Werker, "Learning Innovation Policy based on Historical Experience," *Structural Change and Economic Dynamics* 14, no. 4 (2003): 385–404.

33. B. Oyelaran-Oyeyinka and L. A. Barclay, "Human Capital and Systems of Innovation in African Development," *African Development Review* 16, no. 1 (2004): 115–38.

INDEX